ФОРМИРОВАНИЕ
ПОВЕДЕНИЯ ЖИВОТНЫХ
В
НОРМЕ И ПАТОЛОГИИ

FORMIROVANIE POVEDENIYA ZHIVOTNYKH V NORME I PATOLOGII

ANIMAL BEHAVIOR
Its Normal and Abnormal Development

The
International Behavioral Sciences
Series

editor Joseph Wortis, M.D.

ANIMAL BEHAVIOR
Its Normal and Abnormal Development

by
Prof. L. V. KRUSHINSKII
With a Preface by T. C. Schneirla

Authorized translation from the Russian

CONSULTANTS BUREAU
NEW YORK

The Russian text
was published by
Moscow University Press in 1960.
The translation contains
revisions made by the author
subsequent to the publication
of the Russian original.

Translated by Basil Haigh, M.A., M.B., B.Chir.

Леонид Викторович Крушинский
ФОРМИРОВАНИЕ ПОВЕДЕНИЯ ЖИВОТНЫХ В НОРМЕ
И ПАТОЛОГИИ

Library of Congress Catalog Card Number: 61-15172
Copyright 1962 Consultants Bureau Enterprises, Inc.
227 West 17th St., New York 11, N.Y.

Printed in the United States of America

This book is dedicated
to my mother
Anna Aleksandrovna Krushinskaya

CONTENTS

PREFACE

It is a pleasure to welcome the English translation of the first book by Dr. Krushinskii, head of the Moscow University Laboratory on the Pathophysiology of Higher Nervous Activities, dealing with the theory and principal results of investigations by himself and his associates. The first edition of this book in Russian appeared in 1960.

The author entered science as a secondary school student assisting in the research laboratory at the Moscow Zoo. Then, as an undergraduate student at the University, majoring in genetics, he did his first original research on "phenogenetic characteristics" in dog behavior. In 1938 he served as consultant on the genetics of higher nervous activity in dogs at the Pavlovian Institute of Physiology in Koltushi; during the war he served as scientific advisor on the training and use of military dogs, then was co-author of a book on this subject published in 1952.

In 1948, one year after completing his doctoral thesis on "The Question of Phenogenetics in the Behavior of Animals," Dr. Krushinskii organized at Moscow University the laboratory which he has directed since then. His scientific publications, numbering about 50, have been directed mainly at problems in the physiology, genetics, and pathology of animal behavior.

This report of the author's energetic application of the reflex and related physiological principles to behavior study offers a valuable opportunity to review important aspects of the great radiation of Soviet behavior research stemming from the Pavlovian movement. Clear indications of the background of inspiration for this research appear in earlier developments such as Sechenov's findings in Russia centered on the principle of reflex inhibition and in still earlier Western influences such as the Darwinian revolution and DuBois Reymond's deterministic teachings on living functions.

As a theoretical step toward the principles of behavior, Dr. Krushinskii discusses results bearing on the properties of the passive defense reflex (PDR) and the active defense reflex (ADR) in dogs. His research on the ADR, particularly, represents an appreciable extension beyond the Pavlovian beginnings of such work. Basic to his system are the concepts of: 1) the unconditioned reflex as innate, 2) the unitary reaction which merges in

behavior and 3) the b i o l o g i c a l   f o r m or adaptive behavior system. (In the last term, "form" is clearly meant and not "pattern.")
These concepts constitute a hierarchy which, if seriously considered, makes the plan of his research understandable. It is his
object to show how the complex adaptive behavior of animals comes
about through the integration of "relatively simple innate and individually acquired reflexes" with the influence of endocrine factors.

The author accepts and extends the Pavlovian terms "strong"
and "weak" to characterize the nervous systems of different sub-
jects investigated in terms of the unitary reactions PDR and ADR,
studied, for example, in relation to the problems of training dogs
for specific duties. The readers who may be somewhat disturbed
about the vagueness of these general terms, his answer very probably would be that in his system they are defined operationally and
put to the test in research and in practical training situations.

A particular contribution of Dr. Krushinskii's work is the extension of research and theory on the ADR, considered with the PDR
in relation to the "strength" of the nervous system. He is econom-
ical with such special terms; also, those he adopts are defined
systematically on the basis of behavioral criteria, as are terms
such as "cowardice" and "excitability," and his work is thereby
subject directly to evaluation by readers able to appraise the scope
and worth of the criteria used.

Professor Krushinskii, a geneticist in his early training and
research orientation, is particularly interested in breeding and
training projects with dogs centered on the study of correspondences
between given genotypes and phenotypic behavior, with the PDR and
ADR as his principal tools. For American behavior students this
is "Psychogenetics," and in Chapter 2 some of the principal articles
in the international literature on this subject are discussed. If
statistically minded readers point to the relatively low values of
the best correlations obtained, the author doubtless would indicate
the desirability of also considering the intergroup correlation
trends he has obtained.

Although this research and theory is not specifically develop-
mental, it is developmentally oriented. Thus Dr. Krushinskii, using
the morphine technique in research with selected genotypes and
their hybrids, found results indicating to him that the PDR and ADR
as unitary reactions may be blended in different individuals according to how a given genotype interacts with the conditions of ontogeny
and of testing. It is worth noting that, in recent years, Soviet conditioning research, particularly with respect to functions such as
the "orienting reflex," has increasingly dealt with neonate and early

stages. To be sure, the longitudinal aspects of such behavior have not been the objects of intensive investigation, nor have they been in other countries. Conventional devices such as "instinct" and the Pavlovian "unconditioned reflex," viewed as representing innate, intraneurally determined entities, may bear a considerable amount of the responsibility for this state of affairs.

Professor Krushinskii bases his behavior theory on the unconditioned reflex being innate in the individual. He favors retaining the concepts of the "innate" and "acquired" in behavior, emphasizing that training capacities are improvable both through selection of genotypes and through manipulation of conditions in ontogeny. On the other hand, he maintains that these factors usually interact so closely in development as to oppose a strict dichotomy of the unconditioned and the conditioned. He nevertheless favors an innate—acquired dichotomy as "more precise" than a system based on appropriate concepts of "maturation" and "experience." Dr. Krushinskii makes this statement unequivocally, although his book is not designed to deal with the question of what the criteria of precision may be for the comparison of these opposed dichotomies. In fact, a considerable part of the available relevant evidence is not discussed in these chapters. The possibility exists, for example, that the roles of extrinsic stimulation and of peripheral mechanisms in behavioral development may, from early stages, assume a significance appropriate to the evolutionary level of the given species, and that the part played in such development by secretory factors may well extend beyond that of affecting neural excitability directly.

The fact will be noted that although Dr. Krushinskii seems to favor an intraneural theory of behavioral development, he objects firmly to ethological postulates such as those of "IRM" and its corollary, "vacuum reaction," preferring broader neurophysiological concepts along the Pavlovian trend such as the Ukhtomskii "dominant." It is only fair to mention the fact that although these assumptions of the ethologists lack specific neurophysiological support, the Pavlovian physiological concepts of intraneural processes in conditioning are themselves insufficiently based on evidence.

In the discussions of the role of neural overexcitation processes in behavioral pathology in Chapter 4, and of physiological analyses of animal understanding in Chapter 5, we are reminded that the research program has an evolutionary and phyletic orientation. All research systematically directed at problems of pathological behavior in relation to the properties of species-typical adaptive processes merits attention, especially as theory concerning the role of behavior in evolution is still rudimentary. Readers interested in such problems therefore will make a point of reviewing Chapter 1 in connection with the reading of Chapters 4 and 5 in this book.

In his studies of complex behavior, discussed in Chapter 5, the author widens his survey to include birds and also mammals other than dogs, with the methods extended appropriately for behavioral comparisons. Although, with standard conditioning techniques, animals in different vertebrate classes have been found much alike except in aspects such as rates of extinction, methods designed for the investigation of problem-solution processes in these animals reveal significant differences. Valuable features of this research are its basis in the Pavlovian discipline and its orientation to systematic concepts such as PDR and ADR. A distinctive feature of these exploratory studies of Dr. Krushinskii's on problem-solving behavior, illustrating the extent and nature of current Soviet interest in cybernetics, is represented by the term "extrapolation reflex" for the varied solutions obtained from different animals.

One of the important practical applications of the author's research has concerned the training of animals for communications services. Surely a most important function of his book in this translation, as of all such cooperative arrangements including exchanges of personnel in scientific and cultural fields, must be that of contributing to international communicative processes and understanding.

T. C. Schneirla, Curator
Department of Animal Behavior
The American Museum of Natural History

AUTHOR'S PREFACE TO THE AMERICAN EDITION

This book is devoted to an experimental analysis of some of the complex problems of animal behavior and of pathological reactions of the organism.

All these problems have been extensively studied in America, and the author accordingly hopes that the questions raised herein may be of interest to American investigators.

The investigations carried out by the author and described in this book are based on the reflex concept of the activity of the brain.

In this work the author attempts to show how, as a result of the integration of relatively simple innate and individually acquired reflexes, with the participation of certain glands of internal secretion, the complex hierarchy of the adaptive behavior of animals is built up. Great attention is paid to the problem of the formation of instinctive behavior.

From the standpoint of the reflex theory, an attempt is made to undertake the physiological analysis of the most complex and as yet little-studied forms of animal behavior which may be defined as elementary rational activity. In the book are described the results of experimental work by which it is possible to distinguish a reflex act, called an extrapolation reflex, which evidently is one of the essential components of the physiological mechanisms of the elementary rational activity of animals. The isolation of such a relatively elementary unit of complex adaptive behavior is, we believe, an essential step in the physiological analysis of this type of complex animal behavior.

This section of the book may be of interest for its extension of our idea of the function of the brain as a controlling system in connection with advances in cybernetics, a science originating in America and now widely developed in many of the countries of the world.

The book gives the results of the physiological analysis of certain mechanisms responsible for the marked excitation of the brain in response to the action of a strong external stimulus. The protective measures utilized by the nervous system to withstand excessive excitation of the brain are indicated.

This section of the work is closely related to the investigations of the Canadian pathologist Selye on the role of stress factors in the development of a number of severe pathological conditions of the organism.

Our approach to this problem, however, differs from that of Selye. Whereas Selye defines stress as "the state manifested by a specific syndrome which consists of all the nonspecifically induced changes within a biologic system," attaching predominant importance to the hypophyso-adrenal system in this state, we have concluded from our researches that many pathological reactions (even leading to death) are based upon a state of marked excitation of various structures of the brain. An exceedingly important role in the production of this marked excitation of both the brainstem and the higher divisions of the brain is played by weakening of the inhibitory functions of the nervous system. Under the action of strong stimuli there is a gradual lowering of the threshold of excitability of the nerve centers. As a result the whole mass of hitherto subthreshold stimuli begins to have an exciting influence on the nervous system, maintaining its overexcited state even after the action of the strong stimuli leading to this condition has ceased.

Pathological excitation under the influence of exteroceptive stimuli develops especially readily in individuals with a particular genotype, determining primarily increased excitability of the nervous system.

The main accent in our investigations of pathological reactions has thus been placed on the study of the state of excitability, the process of excitation, and the role of the protective-inhibitory functions of the nervous system in the prevention of pathological conditions of the organism.

In Russia, this path was laid down by the classical investigations of I. M. Sechenov, N. E. Vvedenskii, and I. P. Pavlov.

L. Krushinskii

Moscow, March 15, 1961

FOREWORD

The main purpose of this book is to analyze some aspects of the complex problem of behavior formation in animals in both normal and pathological conditions. There is no doubt that behavior is formed as a result of the integration of very many components, and it is, of course, impossible to take all of them into consideration. At the present stage in the development of the physiology of higher nervous activity, however, we can distinguish the more important components and study their interaction during the formation of behavior in animals in normal and pathological conditions.

The second chapter deals with a study of the role of the excitability and strength of the nervous system in the formation of certain behavioral acts. The material presented in this chapter illustrates an important aspect of Pavlov's teaching concerning the role of the individual properties of the nervous system in animal behavior.

The influence of the glands of internal secretion (the thyroid and the gonads) on the degree of excitability and the strength of the nervous system, and hence on the manifestation and expression of certain behavioral acts, is discussed in the third chapter.

In the fourth chapter we examine the role of a state of over-excitation of the neurons of the brain in the development of various pathological conditions (convulsive fits, motor neurosis, hemorrhagic shock, disturbance of the cerebral circulation, cataleptoid states) and indicate some of the physiological mechanisms used by the brain to combat the state of overexcitation. The material presented in this chapter extends, we believe, the ideas of the Soviet school of physiology (and especially of Vvedenskii and Pavlov) concerning the importance of differences in the functional state of the nervous system in the development of pathological phenomena.

The fifth chapter is devoted to one of the most difficult problems of behavior. In it an attempt is made to distinguish and study a group of reflexes, which we call extrapolation reflexes, based on very rapidly formed associations, by means of which some species of animals are able to grasp elementary cause-and-effect relationships between phenomena in the outside world. We believe that extrapolation reflexes are one of those types of reflex activity of the brain which may be characterized as elementary rational activity.

The identification of a new group of relatively elementary reflexes in the complex behavior of animals may at the same time be
of interest in the mathematical description of the function of the
nervous system, in connection with advances made in cybernetics,
which studies the structure and function of controlling systems of
varied nature. One of the most highly perfected controlling systems
existing in nature comprises the higher divisions of the brain. A
clear understanding of both the elementary and complex manifestations of its activity is therefore necessary for the study of the laws
governing controlling systems.

The material presented herein is based upon the results of
experimental research conducted by the author and his collaborators. The collection of this material was not confined to the laboratory methods of investigation ordinarily used by physiologists. For
many years now the author has participated in the utilization of the
behavior of dogs for practical purposes. This has made possible
the study of distinctive features of the behavior of a large number
of animals under more complex conditions than those associated
with experimental procedures on dogs, and the results are amenable
to statistical analysis. The author's observations on the behavior
of animals in natural conditions of existence have enlarged our
understanding of the complex forms of animal behavior and have
led to further experimental research.

In carrying out the present investigation, the author is indebted
in the first place to Moscow University, where he has worked since
his student days. He also wishes to express his appreciation to the
following:

The laboratory staff – L. N. Molodkina, D. A. Fless, L. P. Dobrokhotova, A. P. Steshenko, Yu. I. Shar, and G. P. Parfenov, who
have assisted with the investigations during the last ten years.

L. G. Voronin, Head of the Department of Physiology of Higher
Nervous Activity, for his attentive interest and for providing research facilities for the work I directed.

The staff of the I. P. Pavlov Physiological Institute in Koltushy,
for enabling me to collect valuable material for this work.

Posthumously, to M. M. Zavadovskii, L. A. Orbeli, M. Ya.
Sereiskii, V. Ya. Pokrovskii, and V. G. Golubev.

G. P. Medvedev and V. V. Rylov, in whose departments the
author carried out part of his investigations.

A. A. Lyapunov, G. I. Polyakov, and Ya. Ya. Roginskii, for their
valuable suggestions and discussions of a number of scientific
problems.

T. A. Bednyakova, for her valuable help in the research during wartime, and for the statistical analysis of the results.

A. A. Semiokhina and M. M. Karimova, for their help in compiling this book.

O. T. Pomalen'kaya, the editor of the book.

CHAPTER 1

THE RELATIONSHIP BETWEEN CONGENITAL AND INDIVIDUALLY ACQUIRED COMPONENTS IN THE FORMATION OF BEHAVIOR IN ANIMALS

The reflex principle of the activity of the nervous system, discovered by Descartes, revived by Prochazka, and brilliantly applied by Sechenov and Pavlov to the explanation of complex forms of behavior of animals and man, is of universal importance. The behavior of animals and man, including its more complex forms, is determined by the influences to which they are subjected in the course of their lives.

In Pavlov's work, which introduced a new chapter into physiology, the principal emphasis was laid, however, not on the study of the laws of behavior itself, but on the study of the mechanism of the reflex activity on which behavior is based.

The study of the laws governing higher nervous activity and its elementary unit—the reflex—demanded special experimental conditions. The principal requirement of such experiments was the creation of conditions in which the reflex response of the nervous system could be most clearly observed. This led to the necessity of excluding the whole varied range of the animal's behavior, which was achieved by keeping the animal under conditions in which it was exposed to a minimal number of external stimuli while the full range of its behavior was restricted. The principal functional indicator of the activity of the nervous system was the salivary reflex. The laws of higher nervous activity which were thus discovered have clearly demonstrated the efficacy of Pavlov's method.

When he established the fundamental laws of higher nervous activity, Pavlov showed that the motor actions of animal behavior are based upon the same laws as were revealed by studies of the salivary reflex. Thus it was shown that the laws of higher nervous activity discovered and studied by means of the "salivary" method are a universal mechanism of nervous activity, lying at the basis of behavior.

According to Pavlov's scheme, the behavior of animals is formed of unconditioned reflexes (the combination of which gives complex unconditioned behavior) and of conditioned reflexes. In these terms the unconditioned reflex activity is defined as congenital, and the conditioned reflex activity as individually acquired. Without such a strict division of reflex activity into congenital and individually acquired, it would be impossible to make progress in the study of higher nervous activity. Food, sexual, defensive, orienting, maternal, and certain other more highly specialized reflexes are, according to the views of Pavlov's school, the congenital basis on which the whole future behavior is built.

When passing from the study of the laws of reflex activity to the study of the laws of behavior itself, however, it is impossible to make such a strict division into conditioned and unconditioned reflexes. Behavioral acts in most cases appear to be the result of the complex integration of conditioned and unconditioned reflexes, which are mingled into single, integral actions. This becomes obvious as soon as the investigator passes from the study of the animal in the laboratory to experimental conditions more natural for the animal.

When a dog standing on a laboratory bench withdraws its paw in response to the shock of an induction current, this is an action which is based on the unconditioned defensive reflex. When the dog withdraws its paw in response to a flash of light previously applied in conjunction with an electric shock, this is a behavioral action based upon the conditioned defensive reflex. When, however, the dog begins to show a passive defensive reaction (cowardice) toward a stranger (runs away, hides, or cringes on the ground), we are confronted by more complex behavior than the actions described above, based upon conditioned or unconditioned reflexes. We cannot classify it as a "conditioned" or as an "unconditioned" reflex.

The principal difficulty is essentially as follows. When we study different behavioral actions in animals, we find that behavior which is similar in its final expression may be brought about by different causes.

For example, the passive defensive behavior of a dog may, on the one hand, be formed as the result of the individual experience of the animal during its training under conditions of isolation (Vyrzhikovskii and Maiorov, 1933). Such training is conducive to the development of cowardice. On the other hand, such behavior appears regularly from generation to generation in certain families of dogs in spite of the fact that they have been brought up in freedom (Krushinskii, 1938). In the overwhelming majority of cases, however, a particular form of behavior is formed by such close interaction between inherited factors and the influence of conditions of rearing that it cannot be classified as either congenital or individually acquired behavior (Krushinskii, 1944, 1946).

The same behavioral acts may thus be formed in some cases as a result of the combination of mainly unconditioned reflexes, and in others under the influence primarily of the conditioned reflex components of behavior. Between these extreme methods of formation of behavioral acts there is a continuous series of gradations, of different quantitative combinations of congenital and individually acquired components. Their presence shows that it is impossible to make a clear-cut division into congenital and individually acquired acts in more complex behavior than that which is based upon the simple reflex.

If the views expressed are correct, then in the study of behavior a difficulty at once arises in the determination of whether we are confronted with unconditioned or conditioned behavioral acts. In its outward expression the behavior may be the same; the paths along which it is formed are different. Behavioral acts which may be formed as a result of different combinations of conditioned and unconditioned reflexes and which at the same time have similar outward expressions, cannot be described either as complex unconditioned or as conditioned reflex acts. We therefore call such behavioral acts "unitary reactions of behavior" (Krushinskii, 1947, 1948).

In introducing this concept we must emphasize that unitary reactions of behavior are single, integral behavioral acts in which conditioned and unconditioned reflexes are combined and integrated. Thus, a whole range of outwardly similar acts performed by animals, each of which is formed by a different combination of conditioned and unconditioned reflexes, is defined by a common term. We consider that this removes the difficulty in the definition of that level of behavior which cannot be completely described by the terms "conditioned" and "unconditioned" reflex.

Thus, the first characteristic feature of the behavioral unit which we have distinguished is that, since it is formed from a combination of conditioned and unconditioned reflexes, in different cases it may be formed by quite different relative proportions of each, while at the same time preserving the same outward expression.

The next characteristic feature of the unitary reaction is its tendency to accomplish a definite behavioral act which may be developed in different ways, but with a specific consistency of final expression.

A dog showing a "passive defensive" reaction and hiding from a man may perform this action differently depending on the conditions in which it finds itself (it may run away, swim away, hide in a corner, etc.); nevertheless, its behavior is characterized by the final execution of this particular reaction—the effort to hide from the object frightening it.

A dog showing an "active defensive" reaction when confronted with a stranger also performs this action differently, for many reasons (the degree to which the reaction itself is expressed, the conditions under which it is performed, the behavior of the object of attack, and so on), but its behavior has a definite pattern of final execution—the attempt to bite the object of attack.

A dog bringing a present or dead game to its master, although it carries out this behavioral act differently depending on a series of conditions, in this case too follows a final fixed behavior pattern— the dog offers the object to its master.

In all these cases we are dealing with a single, integral act of behavior of the animal, directed toward the achievement of a definite and appropriate result.

The question arises why unitary reactions of behavior, formed on the basis of a variable and not strictly determined ratio between conditioned and unconditioned reflexes, may have a similar outward expression and, while varying in accordance with the conditions under which they are performed, nevertheless lead to the achievement of the same appropriate result. We see the reason for this in natural selection, which led to the formation of these behavioral acts.

It seems indisputable to us that in the struggle for existence the important factor is not how a particular behavioral act is carried out, but what contribution it ultimately makes toward the survival of the species. This is the main reason why in evolution, as elements of behavior, reactions were formed which were directed toward the execution of definite, biologically useful acts.

Would these acts be biologically useful if they had a strictly determined and genetically firmly established pattern of execution from start to finish? Undoubtedly they would. We often encounter such behavioral reactions, especially in the lower members of the animal kingdom.

Under the diverse conditions of the external environment, however, those behavioral acts are evidently more useful which, having an identical final pattern of execution, nevertheless do not proceed along a strictly determined path, but which take account of the circumstances in which the animal finds itself. This latter feature is achieved by combining unconditioned and conditioned reflexes into one single reaction.

The unconditioned reflex component of this reaction reflects the result of the species adaptation of preceding generations to the conditions under which the particular species lives; the conditioned reflex component gives these reactions a "vital flexibility," as a result of which the animal can adapt itself to the concrete conditions of existence.

The very close intermingling of the influence of external factors and the hereditary characteristics of the organism which occurs during the formation of unitary reactions of behavior as a result of the integration of conditioned and unconditioned reflexes is due to the fact that these reactions are highly adapted behavioral units, responding both to the demands of the external environment in which the animal lives and to the "historical experience" of preceding generations.

In concluding our description of unitary reactions of behavior we may mention their principal differences from reflexes (whether conditioned or unconditioned). In our view these differences are as follows:

1. The reflex is the simplest integrated unit of activity of the nervous system. The unitary reaction is the simplest integrated unit of behavior.

2. The conditioned reflex is a temporary nervous connection, whereas the unconditioned reflex is permanent. The unitary reaction is a combination of temporary and permanent connections into a single behavioral act.

3. The reflex is carried out in accordance with a definite pattern from beginning to end. The unitary reaction of behavior is characterized by a definite pattern of execution only of its final stage.

4. Reflexes (unconditioned) may take place without the participation of the higher divisions of the nervous system; the unitary reaction, as a result of the incorporation of conditioned reflex components, always involves their participation.

The unitary reaction may thus be defined as: "an integrated behavioral act formed as a result of the integration of conditioned and unconditioned reflexes, the relative proportions of which are not strictly fixed. This behavioral act is directed toward the execution

of a single and appropriate action which, although
performed by different methods, has a definite pattern of final execution."

Having defined the unitary reaction as the elementary unit of
behavior, in which conditioned and unconditioned reflexes are integrated, it is natural to assume that a subsequent degree of integration exists. We consider that unitary reactions are integrated into
categories of behavior which may be designated as "biological
forms of behavior." By this term we mean behavior which, being
constructed from individual unitary reactions, is responsible for
the fundamental vital functioning of the animal.

We accordingly distinguish the following, most general biological
forms of the behavior of animals: 1) feeding, 2) defensive, 3) sexual,
4) parental (the form of behavior connected with care of offspring),
and 5) filial (the form of behavior of the offspring toward the
parents). These biological forms of behavior are most general.
The first four are inherent in all vertebrate animals and also,
evidently, in many invertebrates; the fifth form is less widespread.

Each biological form of behavior incorporates a series of
unitary reactions. For example, the defensive form of behavior is
formed from active and passive defensive reactions. Although each
of these reactions is an independent behavioral act, when combined
they form an integrated, defensive type of behavior (Krushinskii,
1945).

Besides the most general forms of behavior listed above, other
specific biological forms of behavior are found in the members of
different systematic groups. Many vertebrate animals have a complex group of unitary reactions comprising a play form of behavior.
In many species of birds an absolutely specific form of behavior
exists which is constructed of a complex group of varied reactions
pertaining to flight. In beavers another complex group of reactions
is observed; it may be regarded as a constructional form of behavior. Dogs show various forms of hunting behavior which are
so specialized that they cannot be regarded as the feeding type of
behavior with which their origin is connected.

Forms of behavior may be constructed of a combination of
various unitary reactions. For example, the biological form of
behavior connected with care of the offspring is constructed of
unitary reactions concerned not only with the direct care of the
offspring, but also with their defense, the building of nests or
burrows, and so on.

The migrational form of behavior in birds is constructed of a large group of varied unitary reactions. These include behavioral reactions directed toward the collection of individuals into a flock, migration in a definite direction and along definite routes, arrangement in a specific order in relation to each other during flight, the safeguarding of the whole flock by individual members, and so on.

In almost all forms of behavior a constant component is an orienting reaction.

Thus the various biological forms of behavior may have common unitary reactions, which are manifested against different behavioral backgrounds. In order to differentiate between biological forms of behavior and unitary reactions, we may consider the following principal features.

1. The unitary reaction is the result of the integration of individual reflexes. The biological form of behavior is the result of the integration of individual unitary reactions.

2. The unitary reaction is a single-act form of behavior aimed at the execution of a definite, stereotyped, appropriate action. The biological form of behavior is multiaction behavior, frequently connected with prolonged phases of the life cycle of the animal.

The biological form of behavior may thus be defined as: "m u l t i-a c t i o n   b e h a v i o r,   c o n s t r u c t e d   f r o m   i n d i v i d u a l   u n i-t a r y   r e a c t i o n s,   a s s o c i a t e d   w i t h   t h e   s a t i s f a c t i o n   o f   t h e   m o r e   i m p o r t a n t   b i o l o g i c a l   r e q u i r e m e n t s   o f   t h e   a n i m a l."

The object of this chapter is to indicate the paths of formation of unitary reactions and of biological forms of behavior.

Of all the properties of the organism, undoubtedly the most variable is its behavior. The physiological mechanisms lying at the basis of behavior were formed as the result of prolonged evolution, adapting the animal to a wide variety of changes in the external environment. In spite of the enormous importance of external factors in the formation of behavior, the congenital unconditioned reflex foundation of behavior nevertheless is the base on which the animal's behavior is built up under the action of external influences. When the ratio between the congenital and the individually acquired components of behavior is examined (undoubtedly one of the more important problems of the physiology of higher nervous activity), therefore, the study of the degree of inheritance of each particular behavioral act is of great importance. It is only by studying the degree of inheritance of particular behavioral acts that it is possible to proceed to the study of their variation under the influence of particular external agencies. This consideration led Pavlov to organize at Koltushy the first laboratory in the world for the study of the role of heredity in the formation of higher nervous activity.

Several investigations into the role of heredity in the formation of behavior have been published, but they do not bear the character of a systematic study of this problem. It is interesting that hardly any of this research has been utilized in the majority of treatises and monographs on behavior. In our opinion this is explained by the fact that questions of the inheritance of behavior are not always the concern of research workers directly studying problems of behavior, and hence they have not connected the results of the study of the inheritance of the properties of behavior with problems of behavior itself, and have confined themselves to the simple statement that inheritance does occur.

The existing knowledge obtained from the study of the inheritance of behavior may be divided into two main groups: 1) material relating to the inheritance of specific behavioral acts; and 2) material relating to the inheritance of the general properties of nervous activity, and determining the pattern of development of individually acquired skills.

Let us examine the role of heredity in the formation of behavioral acts of the first group, and primarily of defensive behavioral acts in rats and mice, animals in which the work on the study of the genetics of behavior was begun.

Yerkes (1913) studied the inheritance of the complex of malice, savagery, and timidity in rats. He concluded that this complex, typical of wild rats, is an inherited trait.

Coburn (1922) investigated the inheritance of timidity and savagery in mice by crossing wild and laboratory mice. From a large series of cases (1300 specimens) he concluded that savagery and timidity were regularly inherited in these animals. He considered that these properties in mice were determined not by one but by several hereditary factors.

Dawson (1932) also studied the inheritance of savagery in mice. In a large number of cases (3376 animals), by the use of an accurate, objective method, he showed that timidity is inherited in mice as a dominant trait.

Bauer (1956) studied the reaction of aggression in two lines of inbred mice (C57BL/10 and BALB/c). He found that these two lines differ significantly as regards the number of fights arising between the males. Rearing the animals in solitary conditions had no significant effect on the number of fights that developed. The males of both lines, however, showed greater aggression against the males of lines with which they had not been reared.

Research of the same nature was conducted by Sadovnikova-Kol'tsova (1925, 1928, 1931). Setting herself the task of studying the inheritance of properties of behavior, this worker carried out selection of rats by their ability to learn in a maze (Hampton Court). Faster and slower speeds of training could be induced in two lines of rats by inbreeding. The difference between the indices of speed of training (the logarithm of the time taken for ten tests) in the two lines was found to be statistically significant. These results show the genotypic nature of the difference between the training capacity of the two lines of rats. When she analyzed the cause of the prolonged training period of the slow-to-train rats, however, this worker suggested that this may be due to the considerable timidity of the members of this particular family, and not to any slowness of development of conditioned reflexes in them.

In order to solve this particular problem, members of both families were trained in a Stone's apparatus. The characteristic feature of this apparatus is that, as the rat progresses along its course, it is urged on its way by the slamming behind it of a series of doors. The number of mistakes made by the rat in each ten experiments was counted. Thus the rats were trained in this apparatus on the basis of two reflexes: feeding and defensive. It was found that the curve of training in the Stone's apparatus was identical for the two lines of rats. The author concluded from her findings that the difference in the rate of training of the two inbred lines of rats was due not to a difference in their capacity to form conditioned reflexes, but to a difference in their defensive reflex.

The work described above thus was first to show quite clearly that in the formation of properties of behavior such as defensive reactions, hereditary factors play a very important role.

Research conducted on other animals confirmed the findings obtained in rats and mice. Moreover, when describing the role of heredity, most workers point out that the manifestation and expression of the properties of behavior which they studied were dependent upon the conditions of life of the animal.

Humphrey and Warner (1934) came to the conclusion that defensive behavior in dogs is an inherited trait. Fear of strong tactile stimuli and fear of loud, sharp sounds are independent inherited traits. In their manifestation and expression, however, they depend on the dog's previous conditions of life.

Whitney (1947) points out that different breeds of dog react in an extremely different manner to a standard pain stimulus. In a large series of cases he investigated the reaction of puppies of different breeds to a standard injection of antiplague serum. He found that puppies of some breeds (for example, bull terriers) react very weakly to the injection; other breeds of puppies (cocker spaniels) react with a violent passive defensive reaction to the injection.

Interesting results were obtained in wolf-dog hybrids. Stephanitz (1932) claimed that wolf and dog mongrels as a rule were cowardly. Schmid (1940) also described the behavior of hybrids of wolves with dogs and stated that timidity and fear of unfamiliar objects developed in these animals. In the second generation a clear dissociation took place between these signs of behavior. Humphrey and Warner report an attempt to train wolf-dog hybrids. Training proceeded successfully while the animals remained on the leash, but as soon as they were set free they could not be made to obey. Adamez (1930) points out that a tendency toward running wild, which may be associated to some degree with a passive defensive reaction, is most characteristic of hybrids of sheepdogs and wolves, according to the observations of Patagonian shepherds.

Whitney (1929, 1947) investigated the inheritance of trail-baying in hounds. As a result of crossing of hounds which always bayed as they hunted their prey with breeds of dogs that did not bay as they hunted, first generation dogs which did trail-bay were born. In the second generation a dissociation into trail-baying and non-trail-baying dogs was observed.

Whereas bloodhounds as a rule trail-bay, only rare specimens raise their voice when trailing human beings. Whitney admits the possibility of inheritance of this rare peculiarity as a dominant trait. For instance, from a bloodhound which bayed while on the trail of a man, he obtained a litter in which three of the seven dogs bayed while trailing human beings and the other four did not.

Despite the dominant inheritance of the ability to bay, the characteristic intonation of the voice of hounds (usually low, pure, and melodious) is not inherited by hybrids resulting from crossing hounds and other breeds of dog. Whitney points out that the dissociation taking place in the second generation does not suggest a simple inheritance of this particular property.

Marchlewsky (1930) and Whitney (1932, 1947) state that the manner of looking for game with the head erect (the so-called upper scent) is dominant (Marchlewsky) or incompletely dominant

(Whitney). Hybrids obtained by crossing hounds which trail game (therefore with the head down) with companion dogs (seeking game by means of the scent disseminated in the air, therefore holding the head erect) mainly hold their head high, although when seeking game they sometimes trail by scent also.

The manner in which some German setters follow their prey with a bay similar to that of hounds is, according to Marchlewsky, a recessive feature in relation to the silent method of hunting which is characteristic of pointers.

According to Whitney, the bird hunting instinct in setters is dominant over its absence in hounds and other breeds. Darwin long ago drew attention to the tendency of setter dogs to make a stand, i.e., to stop before game. Speaking of the appearance of this peculiarity of behavior, he stated that if one reared dogs which stood for a short time before pouncing on their prey, systematic selection and appropriate training would intensify this feature.

Marchlewsky studied the inheritance of this characteristic and came to the conclusion that the well-marked ability of pointers to stand at point was partially dominant over the less prolonged and steady stand of German setters. He points out that "backing" (a dog pointing not only when aware of the scent of game, but also at the sight of another dog standing in the same way) is exhibited differently in different lines of hunting dogs.

Different breeds of dogs are known to behave differently in relation to water. Breeds such as Newfoundlands and spaniels like to get into water and swim. According to Whitney (1947) this distinctive form of behavior is dominant when spaniels and Newfoundlands are crossed with breeds of dogs having no marked liking for water.

Whitney also made extensive investigations of the inheritance of the distinctive manners in which dogs express their sense of "happiness." Some dogs, when a person shows them affection, characteristically bare their teeth by raising their lips high. Whitney points out that such behavior is shown only toward human beings, and he never observed it to be shown by dogs toward other dogs. Whitney studied the inheritance of this peculiarity of behavior and concluded that it is inherited as a dominant trait, although he adds also that the expression of this reaction depends on the individual experience of the dog.

Keeler and Trimble (1940) investigated the case of inheritance of the highly characteristic behavioral characteristics of Dalmatians. For more than a hundred years on estates in England Dalmatian or "carriage" dogs were selected for their ease of training to run with a carriage. As a result of selection modern Dalmatians are very readily trained to this activity. Individual differences do occur, however, in the position adopted by the dog with respect to the carriage: some dogs run near the front wheels of the carriage, almost beside the hindlimbs of the horses, others run near the back wheels. Keeler and Trimble conclude that the position adopted by the dog is also to some extent determined by heredity.

Research done on other animals, especially birds, confirmed the leading role of heredity in the formation of certain behavioral properties, for example of defensive reactions.

Phillips (1912) crossed different breeds of ducks and concluded that timidity in these birds is determined by the genotype, although the conditions in which they are kept also have a considerable influence.

Leopold (1944) undertook the study of the inherited difference between wild and domestic turkeys. He also investigated hybrid populations which were living freely in some parts of Missouri and which were the offspring of wild turkey cocks and domestic turkey hens. These investigations revealed several differences in the behavior of wild and hybrid populations. Wild turkeys form smaller flocks than hybrids, and wild turkey cocks as a rule keep themselves apart from the females and young birds; this separation is not seen in the hybrid population. Obvious differences also exist in the times of their cry. A characteristic feature of wild turkeys is their ability to detect danger. Species of a wild population fly away when an enemy is still a great distance away; species of a hybrid population allow the approaching foe to come closer, and then fly a short distance away. The behavior of the offspring of wild and hybrid turkeys differs greatly. The young of wild turkeys show a marked tendency toward "concealment," whereas this reaction is appreciably weaker in hybrids and domestic turkeys. Leopold defines "wildness" as an inherited characteristic, enabling both individual species or an entire population to become adapted to natural conditions of existence.

Mazing (1943) investigated the inheritance of photoreactions in Drosophila. Her investigations showed that normal flies possess a positive photoreaction. In lines of white-eyed flies the photoreaction is subdued; this behavioral feature is recessive in relation to the reaction of the red-eyed flies. Flies with vestigial eyes possess a positive, although somewhat slowed, reaction. Vestigial winged flies exhibited the absence of the photoreaction, which is considered a recessive trait in relation to the positive reaction of normal flies.

In other researches Mazing (1945, 1946) showed that the ability of the flies to select a particular nutrient medium on which to lay their eggs was determined to some degree by inherited factors. By selection it was possible to produce lines of flies laying eggs by preference on a medium either with or without sugar.

This group of hereditarily transmitted behavioral properties and propensities for training for certain behavioral actions is of considerable interest. The findings described illustrate how the behavioral acts of animals, complex in their expression, may be brought about by a relatively simple hereditary mechanism.

The defensive reactions of dogs are very convenient for investigation of the role of heredity in behavior formation. These behavioral reactions, often found in dogs of several breeds, are quite clear and constant in their expression and are amenable to quantitative analysis.

The defensive behavior of animals is expressed in two forms: passively and actively defensive.

The passive defensive reaction takes the form of fear of various stimuli and, in particular, as Pavlov pointed out, fear of unfamiliar stimuli. Dogs show a particularly great and constant degree of fear in relation to a stranger. Many dogs, as our observations have shown, which do not exhibit a passive defensive reaction to such strong stimuli as a shot or an explosion give a well-marked passive defensive reaction to a stranger.

The active defensive reaction or, as Pavlov called it, the watch reflex, is also shown to its fullest degree in relation to strangers.

For these reasons we assessed the presence and degree of expression of both active and passive defensive reactions purely in relation to a person strange to the dog. A method of quantitative assessment of each defensive reaction was developed (Krushinskii, 1944, 1947). According to this method, the passive defensive reaction was assessed by a seven point scale: T^0—complete absence of passive defensive reaction, T^1-T^6—increasing degrees of expression of this reaction.

In the dog the passive defensive reaction is shown to its most marked degree usually at the age of one year, and sometimes later. In the adult dog the defensive behavior is constant, maintaining an

identical form of expression for several years (when kept under
ordinary conditions). The coefficient of correlation between two
determinations, made at an interval of 1 to 2 years, was +0.87 ±0.04.

The active defensive reaction in dogs is expressed in two forms.
One is expressed as barking at a stranger, but with no attempt to
bite him; the other includes an attempt to bite. Finally, the dog may
show complete absence of an active defensive reaction in its behavior with human being. With respect to their active defensive
reaction, dogs as a whole may therefore be divided into three groups:

1. Dogs with no active defensive reaction whatever (designated
A^0-aggressiveness absent).

2. Dogs barking at a stranger but not biting (ABa–aggressive,
barking).

3. Dogs which bite (ABi–aggressive, biting).

The last two groups are divided into three subgroups according
to the degree of expression of this sign: $ABa^{1,2,3}$ and $ABi^{1,2,3}$.

The subsequent determination of the degree of expression of
the active defensive reaction over a period of several years showed
a good measure of agreement between the results of the individual
experiments. The value of the coefficient of correlation was
+0.79 ±0.04.

In those cases when it was necessary to have a permanent
record of the active defensive reaction, we recorded on the kymograph the contractions of a rubber balloon placed under the dog's
lower jaw, during barking.

In order to ascertain the role of heredity in the formation of
the passive defensive reaction we investigated two groups of dogs
(Krushinskii, 1938, 1947). The first group consisted of 224 dogs,
mainly German (or Eastern European) shepherds and Airedale
terriers obtained from different crossings. The dogs of this group
were trained in various conditions—by private individuals or in
departmental kennels. The second group consisted of 89 dogs (mainly
mongrels), obtained by crossing dogs from the kennels of the
Institute of Physiology. All the dogs of this group were trained in
the same conditions in the kennels of the Institute in Koltushy.

For the analysis of the results the total number of dogs was
divided conventionally into two alternative groups: those with and
those without the presence of a passive defensive reaction.

The findings relating to the passive defensive reaction presented
in the genealogical diagrams, in the case of the group 1 dogs are
shown by the same principle of alternate division as in the tables,

i.e., the total number of dogs was divided into individuals with or without a passive defensive reaction. The findings relating to the passive defensive reaction of the dogs of group 2 are given in the genealogical tables with a quantitative assessment of the degree of its expression in each individual. Details regarding the inheritance of a passive defensive reaction in the group 1 dogs are given in Table 1 and in the genealogical diagrams (Fig. 1 and 2).

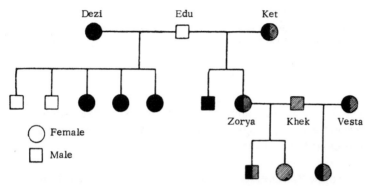

Fig. 1. Inheritance of a passive defensive reaction in dogs. Black symbols— passive defensive reaction; shaded symbols—active defensive reaction; black and shaded symbols—simultaneous presence of active and passive defensive reactions; white symbols—absence of active and passive defensive reactions.

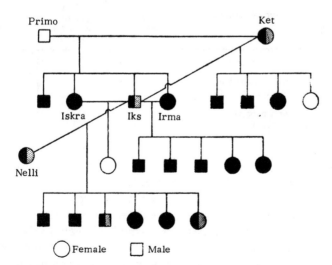

Fig. 2. Inheritance of a passive defensive reaction in dogs.
(Legend as in Fig. 1).

Table 1. Inheritance of a Passive Defensive Reaction
in Group 1 Dogs

Behavior of parents	Number of offspring		
	with a passive defensive reaction	without a passive defensive reaction	Total
Both with a passive defensive reaction...............	45	4	49
One parent with a passive defensive reaction, the other without	61	53	114
Both without a passive defensive reaction	9	52	61

The data on the inheritance of a passive defensive reaction in
the second group of dogs which we investigated are shown in Table 2
and the genealogical diagrams (Figs. 3 and 4).

Table 2. Inheritance of a Passive Defensive Reaction
in Group 2 Dogs (Dogs from the kennels of the
Pavlov Institute of Physiology)

Behavior of parents	Number of offspring		
	with a passive defensive reaction	without a passive defensive reaction	Total
Both with a passive defensive reaction...............	29	1	30
One parent with a passive defensive reaction, the other without	34	6	40
Both without a passive defensive reaction	5	14	19

The details given in the tables and genealogical diagrams indicate the importance of the genotype in the formation of a passive
defensive reaction in the dogs of the two groups. An overwhelming
majority of offspring with a passive defensive reaction was obtained
in those cases when both parents possessed such a reaction. Conversely, a majority of uncowardly offspring was obtained by crossing dogs which did not possess this particular behavioral reaction.

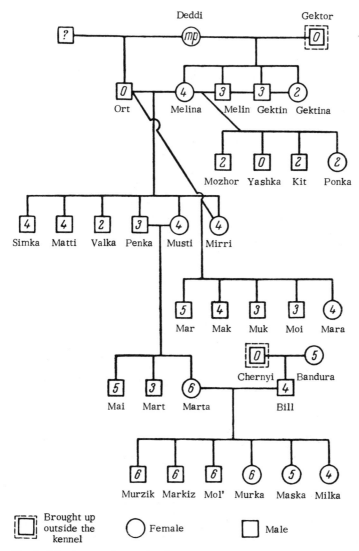

Fig. 3. Inheritance of a passive defensive reaction in dogs (family No. 2) reared in the kennels of the Pavlov Institute of Physiology. The numbers denote the appearance and degree of expression of a passive defensive reaction: 0—absence of reaction; 1-6—increasing degree of expression of a passive defensive reaction; the passive defensive reaction in Deddi and Gektor was not assessed by the author personally.

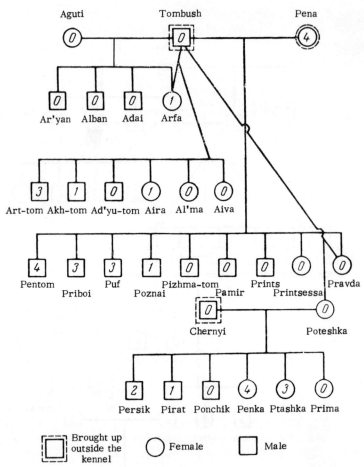

Fig. 4. Inheritance of a passive defensive reaction in dogs (family No. 1), reared in the kennels of the Pavlov Institute of Physiology. (Legend as in Fig. 3)

Comparison of the passive defensive reaction in the dogs of groups 1 and 2 shows that in group 2, in all crossing combinations, more cowardly offspring were found than by the corresponding crossings in group 1. This may be explained by the fact that all the dogs in group 2 were reared in relatively isolated conditions (in the kennel), whereas many of the dogs in group 1 were reared under free conditions (by individuals). In spite of the isolated upbringing of the dogs of group 2, however, in individual families of dogs of this group, for example family No. 1, most dogs either possessed no passive defensive reaction, or it was very ill defined.

To sum up the described findings on the inheritance of a passive defensive reaction in dogs, it may be concluded that congenital components undoubtedly play an important part in the formation of this behavioral reaction. However, the differences in the incidence and degree of manifestation of the passive defensive reaction in dogs reared in different conditions indicate that the mode of upbringing also influences the formation of a passive defensive reaction in dogs.

Table 3. Inheritance of an Active Defensive Reaction
in Dogs

Behavior of parents	Number of offspring		
	with an active defensive reaction	without an active defensive reaction	total
Both with an active defensive reaction.............	21	2	23
One with an active defensive reaction, the other without ..	42	28	70
Both without an active defensive reaction.............	0	28	28

The work of Thorne, who also described the importance of the genotype in the formation of the passive defensive reaction, was published in 1944. A genetic analysis of a group of 178 dogs showed that of 82 displaying cowardice, 43 were the offspring of a very cowardly French basset bitch. Thorne concluded that cowardice in dogs is inherited as a dominant trait, which is unaltered by the individual experience of the dog. Table 3 gives findings illustrating the erroneousness of Thorne's views, which deny the role of the environment in the formation of this particular behavioral act.

Our results show that a passive defensive reaction, although brought about by genotypic factors, is extremely dependent on the training conditions of dogs.

We studied (Krushinskii, 1938) the inheritance of an active defensive reaction in 121 offspring obtained from several crossings of dogs belonging to individuals and departmental kennels (mainly German shepherds and Airedales). The index of the active defensive reaction was the presence or absence in the dog of attempts to bite a stranger. In the genetic data given below, dogs with this tendency to bite are designated as actively defensive.

In Table 3 and Fig. 5 we give the results which we obtained relating to the inheritance of an active defensive reaction.

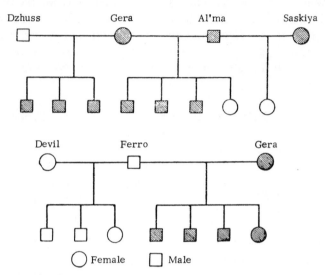

Fig. 5. Inheritance of an active defensive reaction in dogs. Shaded symbols denote the presence of an active defensive reaction; white symbols its absence.

It will be seen from Table 3 that when aggressive dogs were crossed the overwhelming majority of the offspring (21 of 23) possessed the same active defensive reaction as their parents. Offspring from a crossing in which only one parent possessed an active defensive reaction were found either to have or not to have an active defensive reaction. All the offspring obtained from crossing of parents not possessing an active defensive reaction were without this reaction themselves.

It is clear, from the findings described and from reports in the literature, that congenital components play a highly important role in the formation of various behavioral acts of animals. In some cases they play the leading role, and in others they merely create a definite "tendency" which makes it extremely easy to form specific behavioral acts under the dominant influence of individual experience. Congenital components of behavior also undoubtedly are responsible for individual differences in the course of formation of conditioned reflexes. All these considerations indicate the considerable role of heredity in the formation of both the specific and the general features of animal behavior.

Nevertheless, in spite of the considerable role of congenital components in the formation of animal behavior, it is undoubtedly determined very largely by individually acquired experience, by the complete and diverse environment in which the animal lives and is brought up. In the animal's behavioral acts we thus see an intimate mingling and interaction of congenital and individually acquired components.

As Pavlov (1938) pointed out, "the form of behavior of man and animals is determined not only by the natural properties of the nervous system, but also by those influences which have befallen and are continuing to befall the organism during its individual existence, i.e., it depends on constant training or education in the widest sense of these words."* Pavlov later drew attention to the necessity for a careful study of the influence on the animal of all those diverse conditions in which it is reared and in which it lives. Such a study would make it possible to detect both the masked, congenital features of behavior and also the individually acquired ones. On Pavlov's initiative, therefore, his collaborators Vyrzhikovskii and Maiorov (1933) undertook the study of the influence of different conditions of training on the behavior of dogs.

This work was preceded by the investigations of Rozental' (1930), who made observations on the development of puppies and discovered that when they first came into contact with new stimuli, a passive defensive reaction appeared. As they became familiar with the new situation, however, this reaction was gradually inhibited.

Following Rozental's observations, Vyrzhikovskii and Maiorov conducted a special investigation into the isolated training of puppies. Puppies from two litters were divided into two groups. One group was reared in conditions of isolation from the outside world, the other in freedom. Sharp differences were found in the behavior of the two groups of dogs as a result of this training. The dogs brought up in freedom showed no passive defensive reaction, while the dogs brought up in isolated conditions showed such a reaction to a well-marked degree.

This research clearly demonstrated the very important role of the conditions of rearing in the formation of an animal's behavior.

The work of Vyrzhikovskii and Maiorov shed light on an important aspect of the problem of the relationship between the congenital and individually acquired components in the formation of behavior. A part of this general problem is the question of the variability of

*I. P. Pavlov. Twenty Years' Experience of the Objective Study of the Higher Nervous Activity (Behavior) of Animals. Biomedgiz, Moscow-Leningrad, 1938, p. 653.

the instinctive activity of animals under the influence of individually acquired experience. Although this question was raised long ago it still remains unsolved, and moreover different writers express diametrically opposing opinions.

In the opinion of some authorities, congenital or instinctive behavior is not modified by individual experience.

We find a clear exposition of this problem in the writings of Vagner (1913, 1925). "From the facts presented," he writes, "and their number could be increased at will, it follows that instincts are not modified by means of individual adaptation to new conditions of life" (Vagner, 1913). He regards the changes in instinctive behavior observed in captivity as pathological cases, resembling the changes taking place in behavior, for example, after castration.

During recent decades the so-called ethological* trend has developed abroad in the study of animal behavior, the protagonists of which deny that instincts can be modified by the animal's individual experience.

This trend of opinion originated with Whitman (1898), who regarded instincts as congenital reactions which are so constant and characteristic for each species that, like morphological structures (organs), they may be of taxonomic significance.

A similar opinion was held by Heinroth (1918, 1938). By training newly hatched fledglings of many species of birds in isolation from adult birds of their species, he showed that such instinctive movements as preening its feathers, shaking itself, or scratching itself were performed by the young bird without prior experience, in stereotyped fashion, specific for the particular species. If some inborn, instinctive mode of behavior was not characteristic of the bird of a particular species, then it could not acquire that behavior. For example, Heinroth claimed that a species which from birth characteristically does not grasp food with its feet could not learn to do this.

Lorenz (1935, 1937, 1937a, 1939, 1950, 1956) extensively developed this tendency in the study of behavior initiated by Whitman and Heinroth. He also described the extraordinary constancy of instincts, calling them "inherited coordinations, or endogenous movements." Lorenz pointed out that in some taxonomic groups of animals the most characteristic species signs are certain definite instinctive movements. For example, no single morphological sign characterizes the pigeon family so clearly as their "sucking" movements during drinking.

*From the Greek word ethos—character.

Lorenz considers that the most characteristic properties of instinctive acts are as follows. First, their variability is extraordinarily slight, far less than that of the morphological structures of the animal. Second, those instinctive acts which display great constancy in phylogenesis are formed in ontogenesis irrespective of the individual experience of the animal. Third, instinctive movements have a tendency toward spontaneous manifestation even in the absence of specific external stimuli. This manifestation of the instinctive movement through discharge of the IRM (innate releasing mechanism) as a "vacuum reaction" (Leerlaufreaktion) Lorenz has presumed to occur as a result of the accumulation of action-specific energy in the nerve centers, which escapes along the efferent pathways and leads to the manifestation of definite instinctive movements. In the majority of cases, however, in Lorenz's opinion, the manifestation of instinctive acts requires the action of definite and strictly specific external stimuli.

The relationship between the manifestation of congenital reactions and the strictly specific stimuli led Lorenz to the conclusion that a system of specialized congenital liberating nervous mechanisms existed (das angeborene auslösende Schema), which was responsible for the development of reactions in response to definite, strictly specific external stimuli.

Regarding instinctive acts as constant, "organ-like," signs, Lorenz transferred the concepts of experimental embryology into the field of study of the laws of formation of instinctive acts. Just as during the development of morphological anlagen there are definite sensitive periods, in the course of which the anlage of one organ may induce the formation of another, there are also definite sensitive periods in the course of formation of behavioral acts, when they are not yet fully determined and may be modified under the influence of external agencies. In this period, which is characteristic for the formation of every instinctive act, its determination takes place, after which it can no longer be changed throughout life. Lorenz limits the possibility of the influence of external conditions on the formation of the nervous influences lying at the basis of instinctive acts to a definite and very short period in the life of the individual.

Tinbergen (1950, 1955) investigated behavior in a direction close to that of Lorenz, and extended the ideas of the nervous mechanism lying at the basis of instinctive behavior. He considers that a leading role in the mechanism of instincts is played by the

functional activity of the various nervous centers, present in a
definite hierarchic order of precedence. External stimuli causing
the appearance of instinctive behavior do not exert their influence
by a reflex principle (although reflex responses may also be concerned in this process), but merely remove the block hindering
the free passage of impulses from the nerve centers in a state of
readiness to discharge.

The views of Lorenz and Tinbergen thus carry the problem of
the relationship between external and internal in the problem of
the formation of instinctive behavior outside the limits of the reflex
principle of action of the nervous system. Foremost importance
in the development of this form of behavior is attached to the innate,
intrinsic functional activity of the nerve centers. External factors
are relegated to second place in the development of instincts, which
may even appear without a specific external stimulus.

The problem of the variability of instincts under the influence
of individual experience was raised in another form by Lloyd Morgan (1899) in his book "Habit and Instinct." While carefully de-
marcating all that is inherited, or instinctive, from that which is
individually acquired, at the same time he drew attention to the
way both are intertwined in the life of an animal. Instinctive actions
may be modified, as a result of individual experience, to form
behavioral complexes which he called "instinctive habits."

To choose, for example, the change in behavior of falcons when
trained for hunting, Morgan concludes that the individually acquired
experience of the animal is closely connected with its inborn, instinctive behavior. In his opinion it is not only instincts that are
inherited, but also the ability to assimilate individually acquired
experience. Having pointed out that Scandinavian falcons are difficult to train well, and that merlins generally speaking are un-
trainable, Morgan concludes: "They inherit no faculty for responding
to training in this respect. Not only, therefore, is that which is
'congenital' dependent on heredity; that which may be acquired is
also, at all events, limited by heredity." Morgan thus admits the
influence of individual experience on congenital instinctive behavior
and recognizes the presence of continuous transitional forms between instincts and habits.

Bekhterev (1928) also described the variability of instincts
under the influence of individual experience. In his opinion, instincts
are, as it were, an intermediate link between conditioned (conjoined)
and unconditioned reflexes.

Vasil'ev (1941) stressed the role of humoral factors in the development of instincts and pointed out that a purely instinctive behavioral act may be obtained only for the first time during life, and thereafter it must inevitably include a whole series of conditioned reflex components.

Promptov (1937, 1938, 1940, 1945, 1946, 1947, 1956) clearly emphasized the importance and necessity of studying the interaction between congenital and individually acquired components during the formation of behavior. Having carried out many experiments and made numerous observations on the formation of behavior in birds, he found that many stereotyped behavioral acts typical of species of birds are formed as a result of the intimate mingling of congenital and individually acquired components. Whereas in the development of some habits (the character of flight or of movement over the ground or on branches, the ability to pursue prey, the expression of fear, and so on) a leading part is played by the innate basis, common to the species, in the development of other behavioral acts (the character of song of many songbirds, nest building) a more complex combination of natural and individually acquired components of behavior is observed.

On the question of the formation of congenital behavioral acts, Promptov (1956) writes: "Many of the so-called instinctive acts first appearing in the fledgling in the form of a particular motor coordination are initially expressed rudimentarily both in their intensity and in their biological imperfection (and sometimes their irrationality). For example, when three- or four-week-old fledglings (fly-catchers, tits, redtails, etc.) see birds bathing in water, they fly toward them and begin to make bathing movements in some dry place, i.e., near a cup of water: they squat, dip their heads, and shake their wings. Since all these movements take place on the ground, the young bird appears very silly, and its movements are clumsy and biologically irrational. Soon, however, it takes to the water, and the influence of training can then be clearly seen: the young bird does not bathe with the same confidence and energy the first time in its life as the tenth time. It is clear from this example how the developing (and maturing!) inborn coordination is corrected under particular biological conditions, i.e., it is now associated individually with stimuli related to concrete situations.

Very revealing investigations were published by Promptov and by Lukina (1957) showing the role of congenital and individually acquired components in the formation of the character of the song in birds. These investigations showed that birds of different species, when reared from young in isolation, form their song differently, in the manner characteristic of their own species. Some species (lark, tit, peewit) succeed in forming their species-specific song. Others (chaffinch, goldfinch, siskin) when reared in conditions of isolation, formed a song which, although it showed some of the characteristic features of the song of their own species, on the whole differed from it considerably. A third group (starling, speckled magpie) was incapable of forming its species-specific songs when reared in conditions of isolation.

When fledglings of some species of birds were reared with other species of songbirds it was found that the foster-birds to some degree imitated the song of their "teachers." For example, the lark copied the song of the tit and the canary. If, however, the foster-birds were placed in a group of birds which included some of their own species, they always copied the song of the male birds of their own species. Furthermore, if the foster-birds had already formed a song copied from that of another species, they transformed it into the song of their own species.

Intensive investigations along these lines conducted by many foreign workers (Thorpe, 1958; Hinde, 1958; and others) show very clearly the complex intertwining of congenital and individually acquired components in the formation of the character of the song in birds.

Promptov came to the conclusion that "instinctive behavior in birds is the resultant of the complex combination of natural and conditioned reflex reactions."

A new idea of the relationship between innate and acquired was introduced by Schneirla (1956). By studying the formation of behavior among socially developed insects (ants), he showed that a complex relationship existed between morphological, physiological, behavioral, and external factors. In Schneirla's view, during the development of any animal its adaptive behavior is the result of a varying interaction between genotypic factors and the process of ontogenesis, which, in turn, is greatly dependent on external conditions. In animals on any level of phylogenetic development, the organization of behavior is the result of interaction between innate and acquired factors. As Schneirla shows, however, the strictly theoretical terms "innate" and "acquired" cannot be used in the analysis of behavior formation. The term "instinct" therefore loses much of its scientific meaning. Schneirla attaches great importance to the concept of "maturation," characterizing not only the growth of morphological structures, but also the effect of excitation and stimulation of some stages of development by other stages, and the term "experience," characterizing the result of the influence of external factors on development and behavior. The influence of "maturation" and "experience" is closely integrated in the process of ontogenesis. In the opinion of the present author, these concepts are less precise than the terms "innate" and "acquired," which are therefore bound to replace them in the future.

American zoopsychologists such as Lashley (1938) and Beach (1937, 1939), and the behaviorist schools limit the importance of congenital instinctive components in behavior.

An extreme point of view on the question of the role of individual experience in the formation of instinctive behavior was expressed by Kuo (1922, 1932, 1932a, 1932b, 1932c, 1936). After studying the formation of individual reflex reactions in young animals and embryos, Kuo concludes that education in certain reflex acts takes place even during the embryonic period.

On observing chick embryos through a hole made in the shell, Kuo found that three principal reflexes developed—reactions of nodding the head, opening or shutting the beak, and swallowing. He showed that after the eighth day of incubation the embryo developed active movements of the head, accompanied by opening and shutting of the beak, during which time amniotic fluid entered the mouth and was swallowed by the embryo. On the basis of his investigations, Kuo denies the existence of congenital, inherited behavioral reactions, and considers that they are formed entirely as the result of individual experience. This experience may begin even during the embryonic period.

The detailed investigations of Volokhov (1951), however, although they revealed some features of similarity in the general laws of formation of unconditioned motor reflexes in embryos with the conditioned reflexes of adult animals (the phenomenon of generalization and specialization), nevertheless demonstrated the presence of specific regularities of reflex activity formation in embryos of different classes of animals.

The opinions of various research workers which we have described illustrate the enormous divergence of opinion on the roles of the congenital and the individually acquired reaction, and the relationship between them, in the formation of the behavioral acts of animals. As we have mentioned before, this difference of opinion may arise from the fact that different authors have studied different groups of behavioral acts, which are formed with quite different ratios of congenital and individually acquired components. Having defined these integrated behavior complexes as instincts, these authors have reached contradictory conclusions regarding the importance of individually acquired experience in their formation.

Our investigations have shown that not only different behavioral acts, but also the same behavioral reactions, may be formed in different ways. In some cases congenital factors were predominant in their formation, in others—individually acquired characteristics.

Let us consider this question on the basis of concrete examples.

As we have shown above, in the formation of a passive defensive reaction in dogs an essential role may be played by congenital components of behavior. The reaction may also develop, however, under the influence of unfavorable or isolated training conditions (Vyrzhikovskii and Maiorov).

The question arises of the importance of congenital and individually acquired components in the formation of a particular behavioral reaction. To study this problem we investigated the passive defensive reaction in dogs differing by their genotype and by the conditions of their upbringing.

The dogs used for this purpose were German shepherds and Airedale terriers. Individuals of both breeds were reared in different conditions: one group of dogs was reared with individual owners, so that they came into contact with the entire diversity of the outside world; the other group was reared in kennels, largely isolated from the world outside.

Data on the development and degree of expression of the passive defensive reaction in these dogs are given in Table 4.

From Table 4 it is clear first of all that the passive defensive reaction was expressed to a different degree in the German shepherds and Airedale terriers reared in conditions of freedom. Among the German shepherds reared in freedom, the proportion evidencing this reaction was much greater than among the Airedales, and the reaction was stronger in degree in these dogs than in the Airedales. This reflects the congenital mechanism of this particular behavioral reaction of dogs.

Secondly, we see that an isolated upbringing increased both the incidence of the passive defensive reaction and the degree of its expression in both breeds of dogs investigated. The difference was statistically significant in both the German shepherds and the Airedale terriers.

The data in Table 4 also show that the passive defensive reaction developed more acutely and reached a greater degree in the

Table 4. Development and Expression of a Passive Defensive
Reaction in Dogs of Different Breeds, Reared in
Different Conditions

Degree of expression of passive defensive reaction	Airedale terriers				German shepherds			
	in conditions of freedom (individual owners)		in conditions of isolation (kennels)		in conditions of freedom (individual owners)		in conditions of isolation (kennels)	
	No.	%	No.	%	No.	%	No.	%
Absent (T^0)	34	83.0	65	58.5	32	51.5	7	12.0
Very weak (T^1).	3	7.5	21	19.0	6	10.0	1	2.0
Weak (T^2)	4	9.5	16	14.5	14	22.5	11	19.0
Strong (T^3).	–	–	6	5.5	7	11.5	10	17.0
(T^4).	–	–	3	2.5	1	1.5	10	17.0
Very strong (T^5).	–	–	–	–	2	3.0	11	19.0
(T^6).	–	–	–	–	–	–	8	14.0
Total	41	100	111	100	62	100	58	100

German shepherds kept in conditions of isolation than in the Airedales. This was also shown by the fact that the calculated difference
in the degree of expression of the passive defensive reaction in
these breeds kept in isolation was greater than that in conditions
of freedom (and this was also statistically significant).

The fact just mentioned indicates that the influence of isolated
conditions of rearing on the development and degree of expression
of the passive defensive reaction depends on the genotype of the
animal.

In order to pursue the study of this problem, we reared Dober-
man pinschers * in conditions of strict isolation. The puppies (5
individuals), taken from their mother, were placed in an open cage
covered on top with boards. They were cared for by only one person.
Such conditions of upbringing are extraordinarily favorable for the
formation of a passive defensive reaction. However, when these
puppies grew up, they exhibited this reaction to only an insignificant
degree. It was less marked in them than in dogs of other breeds
brought up in far less isolated conditions, and in its degree of
expression it approximated to the passive defensive reaction of the
German shepherds reared in conditions of freedom.

*A breed of dogs in which a passive defensive reaction is hardly found. Among 16 Dober-
man pinschers, kept in conditions of freedom, that were investigated, in 14 we found no
passive defensive reaction whatever, and only in two were minimal signs of this reaction
(T^1) present. In contrast, among the German shepherds reared in the same way, 48.5% of
the species possessed this reaction.

These experiments show that the manifestation of a well-marked passive defensive reaction requires not merely that the dogs should be brought up in strict isolation, but also that there should be a suitable predisposition to the formation of such a behavioral reaction.

The foregoing remarks may be summarized as follows. A passive defensive reaction is formed by interaction between the influences of the genotype and of the external conditions. Under these circumstances the congenital and the individually acquired components of behavior are closely intermingled. In some cases it may be said that this particular behavioral reaction arises under the dominating influence of external factors, and in others, of congenital factors, although such remarks are highly conditional, and applicable only to the most extreme cases. In fact this behavioral act arises as the result of the very close intermingling of congenital and acquired components, integrated into a single reaction.

We showed above that the active defensive reaction of dogs has a congenital basis. What is the role of external factors in the formation of this behavioral act?

Pavlov and Petrova (1916) listed the external conditions in which this reaction develops: "The first is a restricted, or better still an isolated space, in which the dog is present with the experimenter, his master.... The second condition is restriction of the freedom of movement by some type of leash.... Finally, the third condition is the provision of commanding, bold, and varied gestures, of positive and negative character, and actions by the master in relation to the dog in this situation." These conditions are essential for the manifestation of an active defensive reaction already in existence in the dog.

The question arises whether these restrictive conditions for the manifestation of an active defensive reaction also affect the formation of this behavioral act during the individual development of the dog. To study this problem we collected material relating to the influence of isolated training conditions on the formation of an active defensive reaction in dogs of different breeds—in German shepherds and Airedale terriers. The active defensive reactions were compared in dogs reared in conditions of freedom (individual owners) and of partial isolation (kennels). Altogether, 242 dogs were studied.

This investigation showed that in dogs reared in conditions of isolation the incidence of an active defensive reaction is decreased, and its degree of expression is weakened. Both German shepherds and Airedale terriers reared in conditions of partial isolation, were less aggressive than dogs of the same breeds brought up in freedom. Among both breeds of dogs reared in kennels, a considerable proportion of individuals (40-53%) did not show an active defensive reaction; conversely, among the same breeds of dogs reared by individual owners, only 8-16% did not exhibit an active defensive reaction. These results indicate that intercourse between the dog and the varied conditions of the outside world is of essential importance in the formation of an active defensive reaction.

According to Baege (1933), the presence of "rage" (Wut) can be found in puppies from the first day of life. An active defensive reaction in the form in which it appears in the adult dog, however, is found in puppies only after the 52nd day.

According to the investigations of Menzel (1937), if a puppy, 19 days old, is suddenly wakened it may demonstrate an intensive reaction of aggressiveness. Starting with the fifth week, puppies at play displayed a well-marked active defensive reaction in relation to each other. A real reaction of aggression in relation to a human being appeared at the end of the third month. These findings thus suggest the early development of this particular behavioral act in dogs.

For this reaction to develop to its full extent, the animal must have individual experience. Puppies brought up in kennels where the same persons look after them have far less ability to display an active defensive reaction than puppies reared by individual owners.

Isolated training, which is an unfavorable condition for the formation of this behavioral reaction, is at the same time an important and, at times, an essential condition for the manifestation of an active defensive reaction which is already formed, as pointed out by Pavlov and Petrova. The final formation of an active defensive behavioral act thus takes place as a result of interaction between the congenital components of behavior, individual experience, and the situation in which the dog finds itself at a given moment.

According to the work of Scott (1942, 1945), such a picture of the manifestation of active defensive (aggressive) behavior is observed in mice. This writer showed that mice of different lines possess different aggressive tendencies toward each other. This difference is due to heredity. This behavioral act, however, may be considerably modified when the conditions under which the animals are kept are changed. If a mouse of a low-aggressive line is kept with an aggressive mouse and subjected to its continuous attack, the first mouse will also begin to display aggressiveness, and to attack other mice of the low-aggressive line. These findings are in complete agreement with what we have observed in dogs; they illustrate the considerable influence of individual experience on the manifestation of active defensive behavior.

Summing up the results of the study of congenital components and external conditions in the formation of passive and active defensive reactions in dogs, it may be concluded that either reaction is the result of the close interaction between the innate properties of the animal and the influence of the conditions in which the dog lives and is brought up. In this respect, however, considerable differences are observed between them. Whereas intercourse between the dog and the varied conditions of the outside world leads to the weakening of the passive defensive reaction, it also leads, on the other hand, to a strengthening of the active defensive reaction. Figuratively speaking, when the dog during training comes into contact with the diversity of the outside world, it "learns" to be less cowardly and more aggressive. In this respect, as also in relation to the influence of isolated training, active and passive defensive reactions are diametrically opposite.

These findings are of considerable practical importance. They indicate the principal paths which must be followed by those training dogs in kennels for working.

We have shown (Krushinskii, 1952) that in order to obtain a perfectly disciplined working dog it is necessary first of all to make a very strict selection of the progenitors on the basis of their defensive behavioral reactions and, secondly, to have a suitable system of training the puppies. To prevent the formation of a passive defensive reaction and to encourage the formation of an active defensive reaction, it is essential that the puppies should become as familiar as possible with all the varied conditions of the outside world.

Let us examine the relationship between the congenital and individually acquired components of behavior, using the formation of a retrieving reaction in dogs as an example.

With suitable training the skill of seizing and fetching an object thrown or hidden somewhere by the trainer can be developed in a dog. This reaction is widely used by hunting dogs for retrieving game (especially from the water). It is also used by police dogs for bringing the handler objects found on the trail (cartridge cases, and so on). On the basis of this reflex a technique can be built up whereby first-aid dogs, when they find wounded men, announce their discovery to their handlers. They are taught to approach the wounded man, take the man's identification tag in their teeth, and, without dropping it, run to their handler and thus inform him that they have found a wounded man.

We carried out an analysis of the role of congenital and individually acquired components in the formation of this behavioral act (Krushinskii, 1944).

Our first discovery was that the rate of formation of this particular skill in different dogs was extraordinarily variable. We investigated a group of 15 German shepherds trained in the art of retrieving.

The training was undertaken by stages, during which the dogs were taught: 1) to retrieve discarded objects, 2) to take in their teeth an identification tag from around a person's neck, 3) to run to a certain place and, on receiving the command, to take such a tag between their teeth, 4) to run to a certain place and to take the identification tag without command. The dogs thus had to accomplish successively more complicated tasks, all modifications of the basic method, associated with retrieving. Meat was used as unconditioned-reflex reinforcement.

In the process of training great variation was found in the rate of formation of this particular skill. Some dogs consolidated the skill after only 1-2 (in the first stage) to 10-15 (in the fourth stage) performances. Another six dogs could be taught to retrieve objects thrown (first stage) after 5-30 performances, but were quite incapable of learning the method of the fourth stage, or did so only after very many (20-50) performances.

How may this great variation in the ability of the dogs to be trained in this skill be explained? It may be suggested that the dog (whose previous life was unknown to us) had already learned to retrieve. Previous experience, however, can only explain a difference in ability to be trained in the first stage. The techniques of the next three stages are so specialized that no dog lover could teach them to his dog. The existence of parallel trends in the rate of learning in all four stages of training, however, disproves the importance of any previous training in retrieving of the dogs. The presence of inborn properties of the nervous activity in dogs, responsible for their different capacities for training, must be admitted.

On the basis of many years of observation of large numbers of dogs, we have become convinced that individual animals exist which have a constant desire to hold various objects in their mouths. We have found 13 such dogs. Their behavior is very characteristic. Some dogs, when seizing an object and holding it for a few minutes in their mouth, drop it, and then pick it up again a few minutes later. Certain dogs will hold a particular object in their mouth almost constantly (Fig. 6). If the object is taken from the mouth of such a dog, the animal will immediately find and seize another object.

Individual differences in the degree of expression of this characteristic in each dog must be mentioned. Some held small objects in their mouth (pieces of straw, shavings, small chips of wood), others held larger objects (wooden blocks or pieces of iron).

Fig. 6. Manifestation of the retrieving reflex in the dog Ango.

Is the retrieving reaction hereditary or is it the result of individually acquired experience, turning into persistent behavior (of the compulsive motor type) under the influence of various external causes? In order to elucidate this problem we carried out crossing experiments.

It will be seen from the genealogical tables shown (Fig. 7) that in three of the four crosses offspring were produced with well-marked retrieving behavior.

Although the number of cases was small, the results nevertheless suggest that congenital components have an evident role in the formation of this behavioral trait. Among the many hundreds of dogs kept in different kennels, in which we carried out our investigations, this trait was found in only 13 animals. Four of

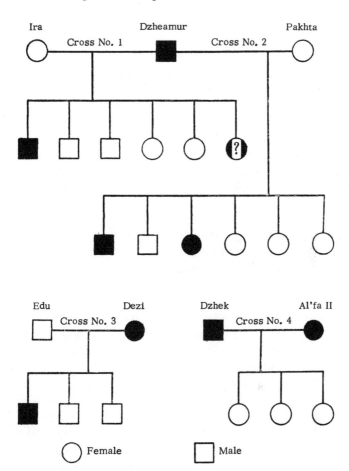

Fig. 7. Inheritance of retrieving behavior. The presence of retrieving behavior is indicated by black symbols, its absence by
white symbols; a puppy in which a retrieving reaction had begun
to appear but which died at the age of 4 months is indicated by a
question mark (?).

these belonged to one family, and two to another. Since this behavioral act is so rarely distributed, the appearance in the offspring of the same behavior present in the parents can hardly
be considered fortuitous. The possibility of imitation of this behavioral act in the offspring produced by Dzheamur is completely
excluded, for his offspring were never in contact with such acts,
and yet well-marked retrieving behavior appeared in three individuals.

The fact that not one of the three offspring of Alf'a II and Dzhek
(dogs which possessed retrieving behavior) showed this property

does not, of course, disprove that it is transmitted hereditarily. It probably indicates simply that its inheritance and the mechanism of its transmission are not simple.

Our results thus suggest that congenital components are evidently involved in the formation of this particular behavioral act. This corresponds to the well known fact that different hunting dogs show different capacities for training as retrievers. Pointers are known to be far more difficult to train as retrievers than continental hounds and spaniels. Among hunting dogs there is even a specially distinguished breed for retrieving killed game—retrievers. The dogs of this breed are very easily taught the art of retrieving, and they have the further ability to avoid crushing the game which they carry. The retriever was obtained by crossing certain breeds of hunting dogs with Labradors.*

The fact that retrievers are more easily trained in retrieving than other breeds of dog (for example terriers) was pointed out by Darwin, who called this property "the natural skill of bringing an object back to its master."

The question arises whether the innate tendency to retrieve facilitates the formation of retrieving skill in a dog being trained. Three dogs possessing this capacity (Chang, Al'fa, Dzhul'bars) were trained in retrieving. Two of them (Dzhul'bars and Chang) began to fetch objects from the first day of training. The third dog (Al'fa) seized objects thrown or placed for it from the first day of training, but only began to fetch them on command after the second day. In later stages of training only two dogs acquired the skill (Al'fa and Dzhul'bars). Both dogs were readily trained in the performance of all four stages. For this purpose Al'fa required ten repetitions, and Dzhul'bars fifteen.

The example given indicates that dogs having the desire to hold objects constantly in their mouth may be taught to retrieve and to perform related acts with the utmost ease.

Is this ready ability to learn retrieving on the part of some dogs a manifestation of some general properties of higher nervous activity, responsible for the rapid formation of conditioned reflexes in general, or is it connected with a special feature of the nervous system—the unconditioned reflex of seizing objects?

In order to elucidate this problem we studied the rate of formation of conditioned reflexes unrelated to retrieving in dogs

*The Labrador is a breed of dog introduced into England at the beginning of the nineteenth century and known to hunters as a good retriever.

having a tendency to hold objects constantly in the mouth. The rate of formation of the conditioned reflexes was investigated in the dog Dzheamur, that showed this tendency, in a room equipped for the study of conditioned reflex activity by the usual salivary secretion method.

These experiments showed that conditioned reflexes were formed at an average rate in the dog Dzheamur. For example, the first conditioned reflex to a metronome with a frequency of 120 per minute (lag 30") began to be formed in the sixth experiment (25-30 combinations), and considerable stability was achieved in the tenth experiment (58-64 combinations). The second conditioned reflex to sound (coincident) was formed after 20 combinations. Differentiation was good. By the type of its nervous activity the dog may be classified as "sanguine" with fairly good cerebral cortical tone. No specific peculiarities were found in the dog's conditioned reflex activity.

In the second dog (Chang), which had this tendency to carry objects and began to retrieve almost without training, a motor conditioned reflex of sitting in response to the command "sit" was formed. This reflex was formed simultaneously in another dog of the same age, reared under the same conditions, but not possessing this tendency to carry objects. As unconditioned stimuli we used pressure with the hand on the buttocks, and food reinforcement when the command was obeyed. In both dogs the conditioned reflex was formed after 25-29 combinations.

Thus in dogs possessing a tendency to retrieve, unrelated conditioned reflexes were formed within normal limits of rate. Conditioned reflexes associated with retrieving were formed exceptionally easily in these dogs.

This capacity of the dogs is thus innate and of an unconditioned reflex nature, and is not the manifestation of the general typological characteristics of the higher nervous activity which determine the rate of formation of any conditioned reflex.

The behavioral act of retrieving, which may be taught to a dog, in some individual animals thus has unconditioned reflex elements leading to its exceptional ease of formation. When this unconditioned reflex is well marked it becomes apparent in the dog's response to certain stimuli without prior training. However, dogs are sometimes found which are not easily trained to retrieve, although training is possible. In these cases training is usually based not on the unconditioned reflex reaction of seizing an object, but on an unconditioned reflex pain stimulus: some weak painful stimulus is applied to the dog, for example pressure with the finger on its lip, and a few seconds later an object is placed into its mouth. The insertion of the object into the mouth coincides with the cessation of the action of the pain stimulus. After a few combinations the dog begins to seize in its mouth an object placed before it on the ground, at first in response to the action of the pain stimulus and later to an appropriate command.

After prolonged training in this way a conditioned seizing reflex, followed by a retrieving behavior, may be formed in practically any dog. The ways of formation of the same behavioral act will, however, differ very considerably in the two cases. In one case it is formed under the dominating influence of an inborn, unconditioned reflex tendency to seize objects, and in the other, by the individual experience of the dog.

Only in these extreme cases can we speak conventionally, in the one case of individually acquired retrieving behavior, and in the other of unconditioned reflex retrieving. In most individual dogs this behavior is formed by means of the closest interaction between the two components of behavior.

The described experimental findings, relating to the formation of defensive and retrieving behavior have shown that each of these behavioral acts may be brought about by various combinations of inborn and individually acquired components of behavior. On the one hand, each may be formed under the dominating influence of individually acquired experience. This is clear from the fact that a passive defensive reaction may develop in an animal when brought up under unfavorable conditions, or from the possibility of retrieving behavior being formed by training a dog in this skill. On the other hand, absolutely identical behavior may be formed under the dominating influence of unconditioned reflex components, as may be seen from the example of the development of a passive defensive reaction in dogs reared in conditions of freedom, although they have an innate predisposition to this form of behavior, or in the manifestation of a well-marked innate tendency to retrieve in certain families of dogs.

Thus the paths of formation of behavioral acts are laid down by the fusion of absolutely different relative proportions of individually acquired and congenital components.

When we speak of a higher or lower relative proportion of innate or unconditioned reflex components of a unitary reaction of behavior, we consider that this is brought about by differences in the threshold of excitability of those nerve centers which are responsible for producing these reflexes. When the unconditioned reflex centers have low excitability, the animal requires prolonged individual experience for the formation of the corresponding unitary reaction. When their excitability is high, the same unitary behavioral reaction is formed even after relatively little individual experience. Finally, if the excitability of the nerve centers is very high, when conditions are created there for the formation of dominant foci of excitation this particular behavioral reaction may develop not only in response to specific unconditioned reflex stimuli, but also to nonspecific, "indifferent" stimuli. A state of "preparedness" to give a definite reaction in response not only to specific, but also to indifferent stimuli in the presence of a dominant was demonstrated by Ukhtomskii in his teachings on dominant reflex components.

We thus consider that the relationship between the degree of excitability of the unconditioned reflex centers and the temporary connections formed through them determines the paths of formation of unitary behavioral reactions. By observing the behavior of an animal we cannot, strictly speaking, answer the question whether this is conditioned or unconditioned reflex in nature. Only in extreme cases, when it is known that its formation takes place under the dominating influence of congenital or individually acquired components, can we speak of conditioned or unconditioned reflex behavior. In most cases it is difficult to determine which of these components is dominant.

All these facts show that between behavioral acts formed under the dominating influence of inborn or individually acquired components there is a continuous series of intermediate degrees with different relative proportions of each, and in many cases this does not permit a sharp line to be drawn between conditioned and unconditioned reflex behavior.

The situation may be summarized as follows. The relationship between the basic units of higher nervous activity (conditioned and unconditioned reflexes) is not strictly determined during the formation of behavioral acts in animals. These may be formed with different relative proportions of inborn and individually acquired reflexes.

Behavioral acts which may be formed with different relative proportions of conditioned and unconditioned reflexes, yet having at the same time an identical outward expression, must be designated by a special term. We call them unitary reactions of behavior. By introducing the concept of the unitary reaction, we describe the general pattern of the relationship between the congenital and the individually acquired components of animal behavior (only, of course, for the level of behavior which is being examined). In accordance with this scheme it is not particular acts of animal behavior that are inherited, but only those unconditioned reflexes from which they are formed by interaction with individually acquired reflexes.

Behavioral acts cannot be innate or individually acquired, but they are formed in the course of the individual life of the animal as a result of the intimate intermingling of innate and individually acquired reflexes. The higher the proportion of unconditioned reflex components in the formation of a particular behavioral act, the more obvious must be the pattern of its inheritance.

The introduction of the concept of the unitary reaction of behavior may remove some of the existing contradictions in the problem of the variability of instinctive behavior under the influence of individually acquired experience. One of the causes of the existing difference of opinion on this problem may be that different writers have studied different complexes of unitary reactions of behavior, formed with very different ratios of innate and individually acquired components. When calling them instincts, they have naturally reached different conclusions regarding the importance of individually acquired experience in their formation.

Let us now pass on to the question of the integration of unitary reactions into biological forms of behavior.

If we examine the complex forms of behavior (which we call biological forms of behavior) by means of which animals adapt themselves to the varied conditions of existence, in several cases we may observe that this multiact behavior is broken down into its separate components, i.e., unitary reactions. Under these circumstances it often appears that the adaptive character of the whole of the animal's behavior is disturbed.

Beach (1937), for instance, caused slight injuries to the cerebral cortex in rats and found that, although as a result of the operation the whole complex of "maternal" behavior in the animals was not abolished, it was, however, broken down into individual, autonomous, disconnected behavioral reactions. The reactions of carrying young rats which had fallen out of the nest, of keeping the young rats clean, and of transporting building material to the nest were preserved. All these reactions, however, while existing separately, did not lead to integrated, adaptive maternal behavior; such parent rats could not rear their offspring. The maternal form of behavior was disintegrated into separate acts, or unitary reactions of behavior.

A similar phenomenon was observed by Promptov (1946) in the nest building behavior of certain birds reared in conditions of isolation. The female chaffinch, reared in isolation, possesses all the separate reactions of nest building, but the order in which they are exhibited is completely disturbed. "The whole of its behavior gives the impression of dissociation of acts which in normal female birds are smoothly connected into a definite chain of successive actions, which are of considerable biological importance in nature" (Promptov, 1946).

A change in the individually acquired components of the normal behavior of an animal, taking place in the one case as a result of trauma to the cerebral cortex and in the other as the result of having been reared in isolation, led to disturbance of the integrity of the biological form of behavior and to its dissociation into individual acts, or unitary reactions of behavior.

The disintegration of behavior and omission of individual unitary reactions is also observed in the sexual form of behavior in hybrid birds. Lorenz (1935) describes a case of such omission of individual reactions in some species and of their manifestation in others. For example a wild duck, when paired with a mongrel drake (wild × domestic duck) was not accompanied by this drake during the period of nest seeking, because this particular reaction was not present in the drake. She was accompanied by an apparently mateless wild drake. However, every time that a quest flight was finished the duck returned to her own mongrel drake. In this case the sexual form of behavior in the mongrel drake was disturbed (omission of the reaction of the quest flight). Meanwhile the quest flight of the solitary duck evoked the appropriate reaction in the mateless wild drake, whose sexual behavior toward the duck was expressed only in the single reaction of accompanying her on her quest flight.

Lorenz cites a similar example dealing with geese. A grey goose was mated with an American gander. During brooding, however, a grey gander without a mate stood "on guard" over her. This reaction was not present in the American gander. After the goslings had hatched, the goose at first went with the American gander. Later, after his protective reaction toward his offspring had died out, the grey gander took his place and began to look after the goslings as if they were his own. In this case certain reactions

characteristic of the grey goose were absent from the sexual form
of behavior of the American gander. The female, however, finding
herself without a mate, evoked in the grey gander the specific form
of reaction which was absent in the American gander.

The sexual form of behavior of the male rabbit can be dissociated into simpler behavioral components (Krushinskii, 1947).
Investigations which we carried out showed that after extirpation
of the whole of the erectile tissue, the penis and urethra (the urinary
bladder was exteriorized to the skin of the abdominal wall), or
after anesthesia of this tissue (by injection of novocain), the reflex
act leading directly to mating was omitted from the behavior of
the male. Such males pursue a female and mount her. However,
having mounted the female, they do not perform the back and forth
movements leading in normal males to copulation. The reaction
to the female, in the form of the behavioral act directed toward
mounting her, was preserved; the reflex act leading to copulation
itself, for which peripheral impulses sent by the erectile tissue of
the penis and urethra were evidently necessary, was omitted. The
sexual form of behavior was dissociated into simpler components.

There is no doubt that these back and forth movements must be
of the nature of a fairly simple unconditioned reflex. So far as the
period before mounting the female is concerned, it is evident that
here an important role is played by the conditioned reflex component of behavior. This was clearly demonstrated in bulls by
Milovanov and Smirnov-Ugryumov (1940).

These examples show that in the sexual form of behavior some
reactions may be omitted and others may be evoked irrespective
of the manifestation of the sexual form of behavior as a whole.

In the above-mentioned analysis of the paths of formation of
active and passive defensive reactions in dogs, we investigated
each of these reactions as an independent behavioral act. Among
dogs, however, we often observe so-called "aggressive-cowardly"
dogs, which simultaneously display passive defensive and active
defensive behavioral reactions. The behavior of the "aggressive-
cowardly" dog is very characteristic. If a strange person approaches
it, it barks at him from a distance, but if the man should approach
closer, the dog runs away, and when far away it again resumes its
barking. This is usually followed by the alternation of one phase
of defensive reaction with the other. In response to sudden movements or to teasing, the dog runs toward the stranger and tries to
bite him, but as a rule fails to do so and runs away with its tail

down. If the stranger tries to go away, the dog pounces on him and tries to bite him. If the person stops and makes a threatening movement, the dog lowers its tail and runs away.

Different explanations are possible for the physiological mechanisms lying at the basis of this peculiar behavior of the dogs. In the first place it may be admitted that there is a single defensive reaction which may be manifested in two phases: one directed toward attack, when the stimulus is not excessively strong (a person a long way off); the other directed toward flight, when the stimulus is excessively strong (a person nearby). In the second place it may be suggested that the behavior of "aggressive-cowardly" dogs is a blending of two autonomous unitary reactions: active and passive defensive, integrated into a more complex, multiact defensive behavior, i.e., into a biological form of behavior. Our investigation showed that the second explanation is right. The defensive behavior of these dogs can be dissociated into separate, unitary reactions of behavior (Krushinskii, 1945).

We found primarily that passive and active defensive reactions are inherited irrespective of each other. The appearance of aggressive-cowardly behavior in dogs takes place as the result of the chance combination of active and passive defensive reactions. Aggressive-cowardly individuals appear when dogs displaying an active defensive reaction are crossed with dogs displaying a passive defensive behavior.

If aggressive-cowardly dogs were crossed with dogs with different forms of defensive behavior (active defensive, passive defensive, aggressive-cowardly, or showing no defensive behavior), the complex of their characteristic defensive behavior was not inherited as an integral item. In each litter obtained from such crosses, individuals were found with different forms of defensive behavior (Figs. 1 and 2). This suggests that the components determining the formation of an active and passive defensive reaction may be inherited independently.

Besides a genetic analysis, we were also able to find a method by means of which the defensive behavior of aggressive-cowardly dogs may be dissociated into its individual components in a physiological experiment. Preliminary experiments showed that the injection of small doses of morphine (morphine hydrochloride, 0.008- 0.04 g) into aggressive dogs leads to considerable suppression of the active defensive reaction, although it does not weaken the passive defensive reaction in cowardly dogs. The effect of morphine on the defensive reactions of dogs becomes apparent after 20-30 minutes, and lasts for several hours. Under these circumstances the general motor activity of the dogs is diminished.

Table 5. Effect of Morphine on Dogs with Different Defensive Behavioral Reactions

No. of dogs	Before injection of morphine		After injection of morphine	
	Defensive reactions	Mean readings of pedometer in 1 hour	Defensive reactions	Mean readings of pedometer in 1 hour
11	absent	676	absent	262
9	passive defensive	2137	no change; increased in one dog	987
12	active defensive	1463	disappeared in 2 dogs; weakened in 10 dogs	232

In assessing the motor activity of the dogs we used a mechanical movement counter, namely a pedometer. The pedometer, which recorded every movement of the dog, was fixed to the animal's neck for one hour before the injection. Twenty to thirty minutes after the injection the pedometer was again fixed to the dog's neck for one hour and a recording of the dog's behavior made. (The dogs were kept in the open cage in which they usually lived.) The results of the experiments are shown in Table 5.

Table 6. Results of the Injection of Morphine into Aggressive-Cowardly Dogs

Name of dog	Pedometer readings in 1 hour		Behavior	
	Before injection	After injection	Active defensive reactions	Passive defensive reactions
Grei	950	300	Weakened	Unchanged
Al'ma	4000	700	Weakened	Unchanged
Roks.	1000	300	Weakened	Unchanged
Dzhek	500	100	Weakened	Unchanged
Nelli.	1300	500	Suppressed	Increased
Amu	1250	300	Weakened	Unchanged
Allyur.	2000	2000	Weakened	Unchanged
Askhari.	700	600	Suppressed	Unchanged
Ledi	500	50	Weakened	Increased
Dzhil'da	1500	50	Weakened	Unchanged
Sil'va	1000	300	Weakened	Unchanged

It can be seen in Table 5 that morphine principally affected dogs possessing an active defensive reaction. In the remaining dogs merely a diminution of the motor activity took place.

If the assumption that different physiological processes lie at the basis of these two defensive reactions is correct, then the injection of morphine into aggressive-cowardly dogs must lead to a weakening, or even to the complete suppression, of their active defensive reaction, without affecting the degree of the passive defensive reaction. The results of the experiments on the injection of morphine into aggressive-cowardly dogs are given in Table 6.

The results in Table 6 show that the injection of morphine dissociates the behavior of the aggressive-cowardly dog into its separate components. It leads to the weakening or even the total suppression of the active defensive reaction, without weakening, and sometimes even with strengthening, of the passive defensive reaction. It will be of interest to describe the behavior of an aggressive-cowardly dog before and after it received an injection of morphine.

The dog Allyur, a male German shepherd.

Behavior Before Injection of Morphine. As I approached the cage (when I was at a distance of 35-40 paces from it) Allyur began to bark and stood by the bars. When I reached the cage the dog still stood by the bars, at a distance of about one pace from me, and barked; 5 seconds later it went away barking to the back of the cage, where it stood and continued to bark. When I struck the bars three times with my hand the dog came running to the front of the cage, but having come as far as the bars it turned back and went into its hut (a closed building connected with the cage). It stood there, poking its head out of the doorway, and growled. When I went away from the cage, with a bark the dog jumped up behind me.

Behavior After Injection of Morphine (0.015 g). As I approached the cage from a distance, Allyur stood silently by the bars. As I came up to the cage, the dog, with its tail down, quietly retreated to the back of the cage; there it stood for a few seconds and went into the hut. When I knocked three times on the bars the dog barked once or twice and hid at the back of the cage. When I went away from the cage the dog did not come out.

The experiments described suggest that morphine differs in its effect on the two components of the apparently unified defensive behavior of the aggressive-cowardly dogs. These facts demonstrate that the behavior of aggressive-cowardly dogs is due to the presence of two defensive reactions which are manifested simultaneously in their behavior. The same stimulus (a stranger) evokes simultaneously two opposite behavioral acts: tendencies toward flight and fight. The behavior of the aggressive-cowardly dog is the result of "interaction" between these two unitary reactions.

The findings described above thus showed that a biological form of behavior could be dissociated into individual unitary reactions. In the case of the defensive behavior of dogs, their behavior is constructed from only two unitary reactions. The more complex biological forms of behavior which are found in animals living in natural conditions, however, are undoubtedly constructed from a much larger number of unitary reactions.

Several investigators have expressed their views regarding the presence of complex biological reactions of behavior. Rozhanskii (1946, 1947) concludes that there is a series of "biological reflexes," which consist of complex groups of simpler reflexes. In contrast to the biological forms of behavior, they are specifically connected with the subcortical part of the brain and the brainstem, i.e., they are complex groups of unconditioned reflexes. Biological forms of behavior, in our meaning of the term, include besides unconditioned reflexes, groups of individually acquired components which take part in the construction of unitary reactions of behavior— those elements from which, in turn, the biological forms of behavior are constructed.

Promptov (1948, 1956) speaks of "biocomplexes of activity" in birds (for example, of the feeding biocomplex of activity), which are single systems of movements, the adaptive integrity of which is not wholly attributable to the reflex mechanism of their component parts. Under these circumstances innate and individually acquired reflexes are united into single, coordinated chains of behavioral acts.

We believe that there is no difference in principle between our "biological forms of behavior" and Promptov's "biocomplexes of activity." From different material and using a different method of analysis, we have reached the same conclusion regarding our ideas of this step in the integration of animal behavior. The term "biological forms of behavior" seems to us to be more suitable than the term "biocomplexes of activity," for a series of unitary reactions forming this category of behavior is characterized not by activity but, on the contrary, by the complete passivity of the animals.

For example, the reaction connected with sitting on eggs is a very important component of the maternal form of behavior in birds, and it is characterized by the complete passivity of the animal. The various unitary reactions which appear in the form of taking cover (at the sight of an enemy, when stalking a prey), forming the corresponding biological forms of behavior, are also characterized by passivity.

The term "biocomplex of activity" emphasizes the necessity of some particular activity of the animal, and behavioral acts such as those we have mentioned would appear to fall outside the group of phenomena characterized by this term.

Tinbergen (1950, 1955), in his system of construction of the various levels of organization of instinctive behavior, used as his starting point the idea first put forward by Craig, and subsequently developed by Lorenz, of the presence of unmotivated, impulsive, appetitive behavior. Appetitive behavior finally becomes consummatory behavior, i.e., the reaction to a definite stimulus.

According to Tinbergen's idea, the nerve center responsible for appetitive behavior in a state of excitation passes under the influence of a series of interoceptive or exteroceptive stimuli and it is induced by excitation of other centers of the nervous system (food, sexual, etc.). A specific stimulus acting upon the animal activates the system of the "innate releasing mechanism," which has until then blocked the possible drainage of excitation from the center, and appetitive behavior begins. When the animal showing this type of behavior meets with the specific stimulus, the block is removed from the centers at a lower level, until the whole chain finally becomes complete in the form of consummatory behavior (the pursuer catches its prey; the male stickleback after migration to the spawning ground, after choosing the site of the nest and building it, and enticing the female into it, fertilizes the eggs).

The whole integrated entity of behavioral acts of animals described by Tinbergen belongs, in our view, in the category which we define as the biological form of behavior. The individual elements of this chain are unitary reactions of behavior. They are all manifested against a background of a particular dominant focus of excitation in the central nervous system. We consider that a physiological basis hardly exists for the separation of individual reactions into appetitive and consummatory. The only difference between these reactions is the order in which they occur. The reaction of catching the prey cannot, of course, precede the reaction of seeking it. Both arise in response to definite external stimuli. The reaction of migration of the stickleback from deep water to shallow and the search for a place to build a nest, which were defined by Tinbergen as "appetitive behavior," result from the action of definite external stimuli: increasing photoperiod and a rise in water temperature. We therefore see no reason for considering those acts of behavior which are defined by these writers as "impulsive" or "appetitive" behavior as reactions differing in principle from other behavioral reactions, and still less do we consider that their mechanism is based upon excitation of special higher centers.

We believe that all these reactions, formed with a different relative proportion of innate and individually acquired components, are integrated into biological forms of behavior, corresponding to the different biological phases of the animal's life.

According to our idea, the biological significance of the fact that complex forms of animal behavior are constructed from individual unitary reactions is as follows. Animals are adapted to the whole diverse range of concrete conditions not by means of biological forms of behavior as a whole, but through individual unitary reactions. In this way it is possible, in the case of changes in the conditions of existence, to reconstruct not the whole biological form of behavior, but only those reactions which are necessary for adaptation to the new conditions of life. The biological form of behavior, however, constructed as a result of integration of individual unitary reactions, is not a simple sum of these reactions. This is seen from the fact that the manifestation of a particular unitary reaction is dependent on the biological form of behavior which dominates the animal's behavior at that moment.

Brüll (1937) investigated the behavior of predatory birds (falcons, hawks), and showed that their behavioral reactions changed according to the territory on which the bird happened to be at a given moment. The behavior of a bird on "nesting territory" (the area around the nest) differed sharply from its behavior on "territory of prey" (an area for hunting). They did not hunt in the nesting territory. Other birds which the predatory hawks would hunt outside this area built nests and fed near the hawk nests.

According to our view, in the nesting territory the maternal form of behavior is dominant in the bird, and therefore stimuli usually evoking a reaction connected with the feeding form of behavior are disinhibited (as a result of negative induction).

On the appearance of an "enemy" (for example, a man), the hawk also reacts quite differently depending on the territory on which he happens to be. If the "enemy" is on hunting territory, the hawk flies away from him, but if he is on nesting territory while the young birds still require feeding, the hawk flies around the "enemy" with a "threatening" cry, and in some cases may even attack him. In this case the same stimulus causes a passive or an active defensive reaction depending on the background of which biological form of behavior the stimulus acts against. When the maternal form of behavior is dominant, as displayed by the behavior of the bird close to the nest with the young, stimuli which usually evoke a passive defensive reaction lead to the manifestation of an active defensive reaction.

These examples illustrate the position which we have expressed, namely that the biological form of behavior, constructed as the result of the integration of a series of unitary reactions, is responsible for their manifestation or inhibition.

We consider that the manifestation of a biological form of behavior is associated with the appearance of a definite focus of a persistent increase in excitability in the central nervous system—a dominant, in the sense in which it was understood by Ukhtomskii (1923). When considering the importance of the principle of the dominant, Ukhtomskii (1945) stressed that the prolonged accumulation of stationary activity of the centers of the nervous system, leading to the lowering of the thresholds of excitability of some reactions and to the inhibition of others, is a mechanism promoting the stability of the animal's behavior in accordance with the principal biological phases of its life.

The biological form of behavior, as we see it, is the external manifestation of the presence of a stable focus of increased excitability in the central divisions of the nervous system, which leads to the easiest possible appearance of certain unitary reactions of behavior and to the inhibition of others. Those unitary reactions of behavior which appear against the background of such a focus of increased excitability form the corresponding biological form of behavior. The focus of excitability itself, however, arising under the influence of both external and internal stimuli, determines which unitary reactions may be manifested at a given moment in the animal's behavior.

When we examine the causes lying at the basis of the development of a particular focus of persistent increase in excitability in the central nervous system, leading to the manifestation of a definite biological form of behavior, we must mention the role of hormonal factors.

The sexual form of behavior, for instance, with the whole range of unitary reactions of which it is composed, reveals itself against a background of an increased concentration of sex hormones in the blood (Zavadovskii, 1922; Beach, 1948; Rosenblatt and Aronson, 1958; and others); the feeding form of behavior arises as a result of hormonal changes in the blood ("starving blood"), creating a focus of persistent increase in excitability in definite divisions of the central nervous system. The maternal form of behavior (in both birds and mammals), with its whole range of constituent unitary reactions, is also dependent on hormonal factors (Ceni, 1927, 1929; Erhardt, 1929; Wiesner and Sheard, 1933; Riddle, 1931, 1935, and others). It is evident that hormones also play an essential role in the formation of behavior of birds. As shown by the work of Vasil'ev (1941), extirpation of the thyroid gland leads to prolonged retention of a characteristic immature behavior in birds.

These facts indicate that the formation of biological forms of behavior is effected with the very close participation of hormonal factors. At the same time, it is hardly possible that individual unitary reactions were formed by the participation of specific hormones. It is only against the background of manifestation of biological forms of behavior that hormonal factors may also influence the individual unitary reactions of behavior.

In the construction of biological forms of behavior from individual unitary reactions a complex interaction thus takes place between the part and the whole. Individual unitary reactions, forming a biological form of behavior, are dependent on the latter for their manifestation. We may therefore rightly regard biological forms of behavior not as the sums of individual unitary reactions, but as entities, as complex levels of integration of behavior.

To conclude this chapter let us pause to consider, from the point of view of the positions which we have adopted, those physiological mechanisms which the protagonists of the ethological school of Lorenz and Tinbergen have placed at the foundation of instinctive behavior. The basic feature in the concept of these authorities is the acceptance of the presence of a system of "innate releasing mechanisms" (blocks), which prevents the drainage of impulses from the excited center until appropriately "triggered." The system may also spontaneously discharge, which leads to the occurrence of a vacuum reaction.

There is no doubt that under the influence of interoceptive and exteroceptive stimuli and of hormonal factors, the excitability of certain centers of the nervous system may be increased. It is, however, difficult to admit that a specific stimulus (for example, food) would cause a flow of impulses along the efferent pathways from the excited center (food) as a result of the removal of the block obstructing it.

As Ukhtomskii showed, if it is possible to produce stationary excitation in a particular central group of a general nervous pathway, which will bring about the lowering of the thresholds of excitability in relation to flowing impulses, the discharge of excitation into the corresponding effectors will be facilitated. In the presence of a dominant in a particular center (for example, in the sexual center) there must therefore occur a lowering of the thresholds for impulses passing from the receptors affected by the specific stimuli (signs of a member of the other sex).

In certain cases, however, these specific stimuli are evidently inadequate. The reaction then takes place only in the presence of additional nonspecific stimuli, as was pointed out by Armstrong (1950), in a critical review of Lorenz's ideas. For example, river gulls in captivity mate with each other only when some unusual object appears. Such facts are difficult to explain from the position of Lorenz and Tinbergen concerning the action of the stimulus in accordance with the "key and lock" principle. Which stimulus here was the "key" removing the block to drainage of excitation from the sexual center?

This phenomenon, however, responds completely to the idea of the dominant. It is obvious that the nonspecific stimulus produced its action as a result of the summation of excitation from the nonspecific stimulus with the excitation of the dominant center. The threshold of excitability of the dominant was inadequately low without this summation to produce a reaction in response to the specific stimulus.

On the other hand, if the excitability of the dominant center is excessively high, its "preparedness" to give a reflex response will determine the possibility of a reaction to a nonspecific stimulus. "This 'preparedness' for a definite reaction, or 'tendency' toward a reaction, decisive in favor of the action of indifferent stimuli, is the expression of a dominant, transferred at a given moment to particular centers" (Ukhtomskii, 1926). The vacuum reaction, described by Lorenz, is evidently the response to some nonspecific stimulus as yet unnoticed by the observer, the excitation from which flows along the efferent pathway of that center in which, at that moment, the dominant has been formed.

Lorenz himself, in one of his more recent works (1956b), draws attention to the importance of "indifferent" stimuli in the manifestation of the vacuum reaction when the threshold of stimulation is extremely low.

We thus consider that the facts described by the protagonists of the ethological trend of opinion may be explained by the physiological laws of activity of the nervous system that are already known; they do not need any hypothesis of a system of blocking mechanisms and of a spontaneous outflow of excitation through ruptured blocks, causing the manifestation of a vacuum reaction.

On the basis of our concept of unitary reactions and of biological forms of behavior it is possible to give a different explanation of the variability of instinctive acts under the influence of individual experience from that given by the supporters of the ethological view.

Since the different unitary reactions and biological forms of behavior, which may be identified with instinctive behavioral acts, are based on differences in the ratios of conditioned and unconditioned reflex components of behavior, the question of whether instincts can vary under the influence of individual experience must be answered in the affirmative.

In cases in which the performance of a particular behavioral act (for example, the drinking of water by pigeons) evidently involves principally unconditioned reflex components, the role of individual experience is insignificant. The examination of such behavioral acts may lead to the conclusion that the formation of instinctive acts of behavior is independent of individual experience.

On the other hand, if we examine the variability of behavioral acts formed with a high ratio of individually acquired components of behavior (for example, during the formation of the song in the males of certain species of imitative birds), it may be concluded that the variability of instincts under the influence of individual experience is extremely great, or even that innate components in general have no part to play in the formation of instinctive acts of behavior.

If, however, the position is accepted (Krushinskii, 1947, 1948) that behavior is always formed from the interaction between innate and individually acquired components, both these conclusions may be considered as correct.

We believe that the notion which we have developed of unitary reactions and biological forms of behavior will make it easier to understand the mutual relationship between innate and individually acquired components in the formation of instincts and habits.

CHAPTER 2

BEHAVIOR FORMATION IN RELATION TO THE FUNDAMENTAL PROPERTIES OF THE NERVOUS SYSTEM

Pavlov showed that, in spite of the wide diversity of animal behavior, certain definite types of nervous activity may be distinguished. "Because our own behavior and the behavior of the higher animals is determined and controlled by the nervous system, it is probable that this diversity of behavior can be reduced to a more or less limited number of fundamental properties of this system, with their combinations and gradations. It thus becomes possible to distinguish types of nervous activity, i.e., complexes of the fundamental properties of the nervous system."*

Investigations conducted in Pavlov's laboratories showed that a difference in the characteristics of the nervous system lies at the basis of the different types of conditioned reflex activity of animals. Nevertheless, very little work has been done thus far on the study of the relationship between the considerable variety of behavioral modalities of animals and the typological features of their nervous system. The resolution of this problem should be of great importance in the establishment of the laws of formation of animal behavior, because one of the possible means of regulating behavior (in both normal and pathological conditions) is undoubtedly through modification of the fundamental properties or modalities of the nervous system which determine the type of nervous activity of the animal.

We have studied the formation of behavioral acts in relation to only two of the fundamental properties of the nervous system: its strength and its degree of excitability. Although these properties of the nervous system do not extend to the other functional properties which characterize its activity (the balance and mobility of the processes of excitation and inhibition), we nevertheless deliberately restricted ourselves to the study of these properties alone. We studied the formation of behavior in relation to the fundamental properties of the nervous system and not in relation to the type of nervous activity for the following reasons.

*I. P. Pavlov, op. cit., p. 651.

First, it appeared to us more appropriate to begin to study behavior formation in relation to the simpler properties of the nervous system, rather than in relation to the type of the higher nervous activity of the animal, which is known to be a complex property.

Second, by using a genetic method for combining different properties of nervous activity, we strove to distinguish those properties of the nervous system which may be most readily inherited. The type of nervous system cannot be inherited intact.

Third, the assessment of the type of nervous activity of each dog by means of those tests which are used in Pavlov's school requires an extremely long time, of the order of months. The individual properties of the nervous system may be studied far more quickly.

Dogs are observed to differ considerably in the degree of their mobility and activity. "Some are reactive, mobile, and sociable to the highest degree, i.e., extremely excitable and quick. Others are just the opposite: reactive, mobile, and sociable to only a very slight degree, i.e., unexcitable and, in general, slow."* This description by Pavlov of the external behavior of dogs may also be applied to other animals. Researchers studying behavior use all manner of methods for recording the movements of animals. Extensive use is made of a revolving wheel, fitted with a counting device, for estimating the motor activity of rats; special cages with a mobile base have been constructed to record the motor activity of dogs, and so on.

Most workers studying the motor activity of animals have used the terms "activity" or "spontaneous activity" to characterize this feature. Although this concept characterizes only the behavior of the animal, it does reflect the manifestation of the general properties of the nervous system, so we have considered that the differences in the motor activity of animals reflect differences in the degree of general excitability of their nervous systems. Animals possessing a highly excitable nervous system and a correspondingly low threshold of stimulation must react to a far greater number of the various stimuli to which the nervous system is subjected than do animals possessing low excitability and a correspondingly higher threshold of stimulation. This is the basis of the difference in motor activity that is observed between individuals of a species.

*I. P. Pavlov, op. cit., p. 657-658.

On the one hand we have extremely excitable animals, characterized by a great deal of motor activity, who react violently to all external stimuli; on the other hand, we have animals of low excitability, characterized by little motor activity. Between these extreme variants there is a continuous series of purely quantitative intermediate forms.

The presence of quantitative differences in the degree of excitability is clearly pointed out by Orbeli (1945). "We may imagine an animal in which weak and strong stimuli evoke weak excitation; we may imagine an animal in which stimulation always evokes excitation of considerable strength, and we may imagine an animal in which the process of excitation is of average strength."*

With regard to the foregoing remarks, by a different degree of excitability we understand a different qualitative e x p r e s s i o n of the functional level of the nervous system, determining the activity of the animal in response to the sum total of external stimuli to which the animal is subjected at each moment of its existence.

By examining the formation of behavior in relation to the general excitability of the nervous system, as assessed by the sum of the movements made by an animal in a definite interval of time, we are fully aware that motor activity alone cannot completely reflect the degree of excitability of the nervous system of the animal being studied. Its excitability may also be manifested along other efferent pathways. Nevertheless, we consider that the motor activity provides the most convenient characterization of the degree of general excitability of the nervous system. This is also confirmed by the fact that in conditioned reflex behavior the motor component is more stable than the secretory.

The same conclusion is reached by Anokhin (1932), who used the method of active selection in dogs. In another of his papers (Anokhin and Strezh, 1933) it is pointed out that the "motor component of the general food reaction, since it is the most highly mobile and has the lowest threshold of excitation, may be present when all the others, especially the vegetative manifestations, are absent."

The differences in the motor activity of animals are largely due to the innate functional properties of the nervous system. Utsurikawa (1917) studied the difference in the behavior of rats of inbred and outbred lines. Recordings of the motor activity showed

*L. A. Orbeli. Lectures on Higher Nervous Activity (Moscow—Leningrad, 1945), p. 96.

that the rats of the outbred line were more active, less savage, and less reactive in relation to sound stimuli. Rundquist (1933) and Brody (1943, 1950) showed that selection can be used with success to isolate lines of rats differing in their motor activity.

The work of Sadovnikova-Kol'tsova (1934, 1938[*]) showed that wild rats possess a higher level of activity than laboratory rats. First generation hybrids obtained by crossing wild and laboratory rats were intermediate between the two parent forms in their activity.

Similar results have been obtained in dogs. Adamets (1930) points out that in Moravia some extremely excitable English pointers, very fast workers during hunting and therefore quickly fatigued, were crossed with less excitable German setters. The hybrids are less excitable than the pointers, but more excitable than the setters.

Anderson (1939, 1941), using the pedometer to assess the excitability of dogs, observes that great differences occur in the daily activity of various breeds. By this property, three groups of dogs were distinguished. The most excitable and active group include German shepherds, Arabian borzois, and spaniels; the group of average activity includes English bulldogs, terriers, and Pekingese; the group of low activity includes French hounds.

Hybrids between the different breeds of animals were obtained. In their excitability, the hybrids of the first generation were intermediate between the extreme parental forms. Most hybrids of the second generation demonstrated approximately the same degree of motor activity as the first generation hybrids, although individuals were found with a very high or, on the contrary, a very low motor activity. This shows that dissociation is taking place.

These facts concerning the inheritance of different degrees of activity in rats and dogs, based on differences in the excitability of the nervous system, demonstrate the undoubted role of innate factors in the development of individual differences in this particular functional property of the animal. It is probable that increased excitability is partially dominant over low excitability.

The importance of the degree of excitability of the nervous system in animal behavior may be elucidated by two methods: 1) by artificially modifying the degree of excitability of the nervous system, with the subsequent study of the change in behavior; and 2) by comparing the behavior of animals with different degrees of excitability. Most authors have studied this problem by establishing

*See under Kol'tsova, M. P., 1938, in the bibliography.

a relationship between the degree of excitability of animals and the course of formation of individually acquired habits.

In Pavlov's school extensive use has been made of drugs which modify the excitability of the nervous system. In this respect caffeine has been most thoroughly studied. The study of the effect of caffeine on higher nervous activity began with the investigations of Nikiforovskii (1910, 1911), who showed that the use of this preparation in small doses leads to an increase in excitability. In his opinion inhibitory processes are not affected by caffeine. The result of strengthening of the process of excitation is an increase in the intensity of conditioned reflex responses and the disinhibition of differentiation.

It was subsequently discovered that caffeine strengthens the process of excitation (Zimkin, 1928; Lindberg, 1935), and moreover that the degree of strengthening depends both on the dose of the drug given and on the state of the animal's nervous system. It was shown, first, that excessively large doses of caffeine, causing a sharp increase in the process of excitation, produce limiting inhibition (Lindberg, 1935; Kleshchov, 1938; Zeval'd, 1938). Consequently, the conditioned reflexes are not increased; on the contrary, they are depressed.

Second, it was shown that the effect of caffeine on the process of excitation is largely dependent on the typological properties of the nervous system. In animals with a weak type of nervous system, administration of only small doses of the drug leads to an increase of the conditioned reflexes; larger doses depress conditioned reflex activity (Kleshchov, 1938). Just as during weakening of the nervous system by excessive nervous strain, castration (Petrova, 1936), or old age (Pavlova, A. M., 1938) only the slight increase in excitability obtained by the use of small or average doses of caffeine leads to an increase in the conditioned reflexes. The administration of large doses depressed them.

The study of other drugs increasing the excitability of the nervous system — strychnine (Nikiforovskii, 1910; Zhuravlev, 1938; Pyshkina, 1939) and cocaine (Zhuravlev, 1938) — showed that these drugs increase both conditioned and unconditioned reflexes.

Investigations devoted to the study of the influence of endocrine factors on the course of formation of conditioned reflexes gave similar results. Research carried out in Pavlov's school by Val'kov (1925), Petrova (1936), Zeval'd (1947), and others, and also experiments undertaken on the initiative of Stockard (1941) and performed by Anderson (1941), showed that extirpation of the hypophysis, the thyroid and parathyroid glands and the adrenals leads to a considerable decrease in the magnitude of conditioned reflexes. Injection of the hormones of the anterior lobe of the hypophysis or of the parathyroid and thyroid glands and of adrenalin leads to an increase in the conditioned reflexes. When analyzing the mechanism of action of these hormones on conditioned reflex activity, certain authors (for example Anderson) found direct indication that hormones act by modifying the excitability of the nervous system of the animal.

Research in which drugs modifying the excitability of the nervous system were used thus showed that the conditioned and unconditioned reflex activity of the animal is dependent on the degree of excitability of the nervous system. An increase in excitability within limits not causing limiting inhibition leads to the strengthening of both conditioned and unconditioned reflex activity.

The study of the relationship between the degree of excitability of the nervous system (assessed on the basis of the motor activity of the animal) and the capacity for development of individually acquired habits gave similar results. The researches of Shyrley (1928), Ligon (1929), and of Tootle and Dykeshorn (cited by Anderson and Smith, 1932) showed high coefficients of correlation (+0.40 to +0.63) between the values of the motor activity and the rate of formation of conditioned reflexes of walking through a maze. Rundquist and Heron (1935), investigating the 17th generation of a line of rats selected in accordance with the sign of high motor activity, came to the conclusion that excitable, active animals learn to pass through a maze more readily than those with a low level of excitability.

Humphrey and Warner (1934) investigated the ability to respond to training of a large number of dogs in the kennels of the dog-breeding service in Switzerland and found that those which learned best were the highly excitable animals. Of 254 highly trained dogs, 70% were of a high level of excitability, and of the 96 which responded least satisfactorily to training, only 33% were of a high level of excitability. The coefficient of correlation between excitability and training ability was +0.40 ±0.03.

Our investigations on 271 dogs also showed a clear correlation between the degree of excitability and the successful development of a motor conditioned reflex. The excitability of the dogs was measured by means of a pedometer. Investigations were carried out on hungry animals (at the time of their evening feeding).

The dog was tied by a 1.5-meter lead to a stake driven into the ground. A pedometer was suspended from the dog's neck. The master (the person who regularly looked after the dog) allowed the dog to lick from a bowl of food he was holding and then, summoning the dog, ran 10-15 meters away, and for the next two minutes continued to show the dog the food and to summon it. All the dogs were in a more or less excited state: they rushed about, raged at their master, and most of them barked excitedly. These movements were assessed collectively by the pedometer. A considerable variation was observed in the intensity of motor excitation of the dogs, which ranged from 10 to 360 movements in two minutes.

During wartime, test dogs were trained for the antitank or communications services. In the dogs of the antitank service the conditioned reflex to run toward an approaching tank was formed. The AT1 dogs got between the caterpillars of the tank; the AT2 dogs, having run up to the tank, ran alongside it. The result of formation of the conditioned reflexes was evaluated throughout the entire period of training of the dog. The quality of the dogs' work was assessed by a four-point system (excellent, good, fair, and poor). The coefficients of correlation between the degree of excitability (measured by the pedometer) and the quality of the dogs' work are given in Table 7.

Table 7. Correlation between the Degree of Excitability of Dogs and Success in Formation of the Motor Conditioned Reflexes

Group No.	n	Coefficient of correlation $r \pm m_r$	Reliability of correlation r/m_r
2 (communications) . .	56	+0.43 ±0.10	4.30
2 (AT1)	100	+0.33 ±0.09	3.67
3 (AT1)	54	+0.52 ±0.10	5.20
4 (AT2)	61	+0.58 ±0.08	7.25

Table 7 shows the presence of a positive relationship between the degree of excitability of the dogs and the success achieved in the formation of the individually acquired habit. The highly excitable dogs formed the individually acquired habits better than did the dogs with low excitability. This is seen especially clearly from a comparison of the results of training of these two groups of dogs (Table 8).

The data given in Table 8 illustrate the difference in the formation of individually acquired behavior in dogs of high and low excitability. Among the highly excitable dogs a high proportion did

Table 8. Success in Formation of Individually Acquired Habits in
Dogs of High and Low Excitability

Evaluation of success in formation of habit	Highly excitable dogs (giving more than 300 pedometer beats per 2 min)						Dogs of low excitability (giving less than 60 pedometer beats per 2 min)					
	Group No. 1 (communications)	Group No. 2 (AT1)	Group No. 3 (AT1)	Group No. 4 (AT2)	Total	%	Group No. 1 (communications)	Group No. 2 (AT1)	Group No. 3 (AT1)	Group No. 4 (AT2)	Total	%
Excellent...	2	3	2	6	13	34.5	0	0	0	0	0	0
Good......	6	8	0	4	18	47.5	0	2	0	0	2	9.0
Fair	0	5	0	0	5	13.0	1	2	3	0	6	27.0
Poor......	1	1	0	0	2	5.0	2	6	4	2	14	64.0

their work well or excellently; on the other hand, among the dogs of low excitability the overwhelming majority worked poorly. Of the 22 dogs of low excitability not one was evaluated as "excellent," and only two dogs as "good."

When the quality of the work of the trained dogs was being assessed, factors of importance were the tempo and the precision of performance of the habit, as well as the ratio of correct performances to failures. The precision of performance of the task is especially important in the case of the antitank job AT1 when the dog, having run up to the moving tank, must go in between its caterpillars without any delay. In order to penetrate under the tank without delay the dog must inhibit its natural fear of such an extremely strong stimulus as a tank moving toward it at high speed.

It was to be expected that the degree of excitability of the dogs, as we assessed it by means of the pedometer, would be responsible for variations in the tempo of performance of the task. Our analysis showed, however, that not only the speed at which the dog ran to its objective, but also the precision of the reaction of penetrating beneath the tank (as a result of the ease with which the dog inhibited its fear of the tank), was in clear correlation with the degree of its excitability (coefficients of correlation $+0.55 \pm 0.10$ and $+0.57 \pm 0.09$, respectively).

Having found a positive relationship between high excitability and the trainability of antitank dogs, we carried out experiments to ascertain the possibility of improving the results of training of the dogs receiving poor marks by artificially increasing the excitability of their nervous systems. For this purpose we used caffeine sodium benzoate and thyroidin. Experiments were carried out on eight antitank dogs, which had responded poorly to training, worked slowly, and which often failed to penetrate under the tank (Table 9).

Table 9. The Effect of Artificial Increase in Excitability on the Trainability of AT1 Antitank Dogs

Serial No.	Name of dog	Evaluation of dog's work before beginning of administration of drug	Daily dose of drug (grams)	Duration of experiment (days)	Dose of drug received during experiment	Evaluation of dog's work at end of experiment
1	Tobo	3	Thyroidin 5—10 Caffeine 0.5	5	Thyroidin 30 Caffeine 2,5	5
2	Orel	2	Thyroidin 5 Caffeine 1	8	Thyroidin 40 Caffeine 8	5
3	Marta	2	Thyroidin 5 Caffeine 0 5	4	Thyroidin 10 Caffeine 2	2
4	Tinta	2	Thyroidin 5—15.0	7	Thyroidin 60	2
5	Daik	2	Thyroidin 5	6	Thyroidin 30	5
6	Rem	2	Thyroidin 5—10	17	Thyroidin 80	2+
7	Tril'bi	2+	Thyroidin 5	11	Thyroidin 55	3+
8	Ema	2	Thyroidin 5—30 Caffeine 1—1.5	31	Thyroidin 355 Caffeine 11	5

It is clear from Table 9 that the administration of thyroidin and caffeine in some cases considerably improved the performance of the individually acquired habit. This effect, however, was not always observed. In some cases (dogs Nos. 1, 2, and 5) the dog's work improved very quickly. On the 5th-6th day of administration of thyroidin and caffeine, the dogs neatly and with great "interest" began to perform the tasks required of them. In other cases (dog No. 8) the effect developed slowly. The lack of effect in certain cases (dogs Nos. 3 and 4) was probably due to the short duration of the experiment (the dogs were withdrawn for unrelated reasons).

To provide a clearer illustration of the influence of these drugs on the performance of this task, we give below a more detailed account of an experiment with one of the dogs.

Ema is a German shepherd, aged 5-6 years. Before the beginning of our experiment the dog had been trained in the antitank duties for two months. The working speed of the dog was extremely slow: the dog approached the tank one step at a time, and in most cases refused to go underneath it; training had been abandoned.

Only nine days after administration of thyroidin to this dog, it manifested signs of increased excitability during work. It began to bark, snarl, and growl at the tank. Subsequent simultaneous administration of thyroidin and caffeine led to a further improvement in the dog's work, although it did not receive an evaluation of "excellent" until the 21st day of administration of the drugs. Continued administration of the drugs led to precise and willing work, performed rapidly. The dog was transferred to an active unit. After eight months (without administration of drugs) the dog still repeated the established pattern, although the speed of performance was now much slower and its work was assessed as only "good."

These investigations thus showed that increased excitability of the nervous system is a condition favoring the development in dogs of individually acquired habits, in the performance of which importance is attached both to the speed of working and to inhibition of the fear of the stimulus in response to which this particular conditioned reflex is formed. Increased excitability in dogs, however, is not of positive importance in all forms of formation of complex motor conditioned reflexes.

Jointly with Fless, we conducted an investigation to ascertain the importance of the degree of excitability of trailer dogs. The principal habit in which these dogs are trained is "trailing work." The dog is taught by command to follow the trail of a person whose scent it has identified by smelling around a place where the person has been or by smelling an object which the person has left behind. The animal must be able to differentiate this scent from that of any other person who has crossed the trail.*

The excitability of the dogs was determined with a pedometer, by the technique described above. The investigation was carried out on 102 dogs trained as trailer dogs. It was found that no obvious relationship existed between the degree of motor excitability and the trainability of the dogs.

In some cases dogs of low excitability excelled at trailing work. For instance, a dog of low excitability (Reks, pedometer reading

*The method of training is described in "The Work Dog" (Sel'khozgiz, 1952).

35) followed a trail at a slow rate, but kept its nose carefully to the scent and nicely negotiated the corners and turns of the trail. This dog possessed a first-class sense of smell.* Reks was one of the best tracker dogs in the group.

Highly excitable dogs engaged in trailing work usually performed it with interest and at a high speed, but often jumped the corners and turns of the trail or switched to another trail. The very excitable dog Reid (pedometer reading 273), for instance, who had an excellent sense of smell, was extremely excited when trailing and, consequently, ran over the corners of a trail, which led to frequent loss of the trail.

Highly excitable dogs are "difficult" animals to train in trailing work, but in the hands of experienced trainers they give excellent results.

Since the training of dogs in antitank and communications work was built up on the basis of an unconditioned food reflex, it might be supposed that the motor activity demonstrated by the dogs excited by the exhibition of food mainly reflects the food excitability of the dogs. The trainability for antitank and communications work, of course, depends on this excitability. However, since trailing behavior was built up on the basis of a different reflex—the active defensive reflex—such a relationship could not be determined. In one group of tracker dogs the motor excitability was therefore determined not only by means of excitation of the dogs with food, but also by excitation at the sight of a stranger. In this case a dog, wearing a pedometer on its neck and tied to a peg, was teased for 10 seconds by someone unfamiliar to it, who made threatening movements directly in front of the dog's muzzle and then retreated to a distance of 20-25 paces (in full view of the dog) and stood motionless for one minute. The dog's master then took off the pedometer. Besides recording the number of movements made by the dog in that minute, the experimenter also noted the time during which motor excitation persisted in the animal (after teasing had stopped).

Comparison of the results of determination of the motor excitability of dogs, in response to excitation by food and by a stranger (i.e., "food" and "active defensive" excitability) showed an obvious similarity in the number of movements made by the dog when excited by the different methods (Table 10).

Table 10 shows the resemblance in the indices of motor excitation in dogs as elicited by the different methods. Dogs with high

*The sense of smell was determined by an objective method developed by ourselves (Krushinskii, Chuvaev, and Volkind, 1946).

Table 10. "Food" and "Active Defensive" Motor Excitability in
Extreme Cases of Investigated Dogs

| Name of dog | "Food" excitability (mean of three pedometer readings over 2 min) | "Active defensive" excitability | | Notes |
		Mean of three pedometer readings over 1 min 10 sec	Duration of motor excitation (in sec)	
Dzhek	281	163	over 60	Dogs with highest
Dzhul'bars	278	177	over 60	pedometer readings
Reid	273	166	over 60	during determination of "food" excitability
Reks I	45	55	3	Dogs with lowest
Reks II	35	33	1	pedometer readings
Al'ba	19	9	0	during determination of "food" excitability

"food" excitability also possess high "active defensive" excitability. This indicates that the difference in motor activity which is detected by exciting dogs with food is not entirely the result of their different food excitability; it reflects a more general functional property of the nervous system.

No relationship could be established between the trainability of dogs in trailing work and the degree of their motor excitability detected during the manifestation of an active defensive reaction. Dogs with different "active defensive" excitability were trained in trailing work with equal success.

Consequently, although trailing work is constructed on a basis of an active defensive reflex, differences in the degree of manifestation of the motor component of this reaction play no significant role. The formation of different motor habits or skills in dogs thus bears different relationships to the degree of excitability of their nervous system. Habits such as are acquired in training for antitank and communications work, in which the basic success of the dog's work depends on the formation of a definite, positive conditioned reflex (running to the tank in the case of antitank dogs), thus bear a clear, positive relationship to the degree of excitability of the dogs: in highly excitable individuals these conditioned reflexes are formed more readily.

Habits such as trailing work, in which great importance is attached to the precise differentiation of stimuli, bear no clear, positive relationship to the degree of excitability of the dogs. In highly excitable dogs, because of the disturbed differentiation, the quality of trailing work is inferior. This is undoubtedly the reason why during training in this procedure no positive relationship can be established with the degree of excitability of the nervous system.

The importance of the general level of excitability for the manifestation and expression of unitary reactions of behavior was clearly shown in our investigations of formation of a passive defensive reaction in dogs (Krushinskii, 1938). The grounds for starting this research were provided by the following fact. Three male Huskies, not possessing a passive defensive reaction, were crossed with German shepherds, also not possessing this particular behavioral reaction. The 25 puppies from these crossings possessed an extremely well-marked passive defensive reaction (Fig. 8).

It is difficult to explain the presence of this behavioral reaction purely on the basis that the hybrids were reared in isolated conditions (in a kennel nursery), for some dogs of other breeds (German shepherds in particular) reared in the same kennels either possessed no passive defensive reaction whatsoever, or it was less pronounced.

We observed a similar manifestation of a well-marked passive defensive reaction in hybrids of wolf and dog found in the Moscow Zoological Gardens and described by Il'in (1934). The hybrids were far more cowardly than wolves.

The presence of a well-marked passive defensive reaction in both cases of hybrids led us to consider that we were dealing with a phenomenon of the same order. In the case of wolf-dog hybrids there was every reason to suppose that the passive defensive reaction was inherited by them from the wolves (since wolves possess it) and was intensified as a result of hybridization.

When both types of crossings are compared, attention is drawn to the low motor excitability of both the wolves* and the Huskies, especially when compared with many breeds of dogs. In the case of the crossing of wolves with dogs, we explain the well-marked passive defensive reaction in the hybrids by the fact that they inherit from the wolves the reflex basis for the manifestation of the passive defensive reaction, and from the dogs the increased excitability of the nervous system. With this combination there is an increase in the degree of expression of the passive defensive reac-

*The chaotic motor excitability which characterizes many breeds of dogs is, as a rule, absent in wolves.

tion, which is the obvious reason for its greater prominence in hybrids than in wolves.

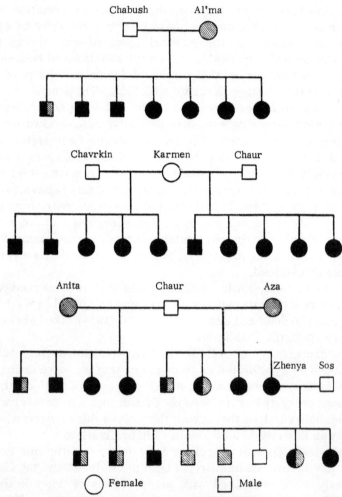

Fig. 8. Manifestation of a passive defensive reaction in crosses of German shepherds (Al'ma, Karmen, Anita, Aza) with Huskies (Chabush, Chaur, Chavrkin). The presence of a passive defensive reaction is shown in black; an active defensive reaction, by shading.

The use of this scheme to explain the development of a well-marked passive defensive reaction in hybrids (of Huskies and German shepherds) meets with a difficulty, for the dogs of neither breed exhibit this particular reaction. We postulated that Huskies have an innate unconditioned reflex basis for the formation of a passive defensive reaction, which is not manifested, however, because of the excessively low level of excitability of their nervous system. When these dogs are crossed with highly excitable shepherds, the same combination takes place as when wolves are crossed with dogs: a combination of the innate basis for the formation of a passive defensive reaction with the increased excitability, which leads to the appearance of well-marked cowardice in the hybrids.

Let us denote the genotypic basis of the passive defensive reaction by T, its absence by t, increased excitability by E and low excitability by e. Upon crossing, we shall then have:

Wolves with dogs		Huskies with shepherds	
Wolf × dog		Husky × shepherd	
eeTT × EEtt		eeTT × EEtt	
Low excitability	High excitability	Low excitability	High excitability
Insignificant cowardice	Absence of cowardice	Absence of cowardice	Absence of cowardice
Hybrids		Hybrids	
EeTt		EeTt	
Increased excitability		Increased excitability	
Well-marked cowardice		Well-marked cowardice	

The principal assumption made in this scheme is that the formation of the passive defensive reaction is dependent on the degree of excitability of the nervous system; if excitability is low, no such reaction may be manifested.

When this working hypothesis was put to the test, several difficulties were encountered. The greatest difficulty was the determination of the excitability in hybrids by comparison with their sires—Huskies. The passive defensive reaction in the hybrids was so well marked that any form of laboratory experiment with them was practically impossible. The presence of an unfamiliar person or a new set of circumstances evoked in these dogs a well-marked passive defensive reaction (the animal hid in a corner and would not react to any stimulus). However, in accustomed surroundings, e.g., in the open cages where the hybrids lived, and in the presence of their regular attendant, the dogs displayed all the signs of increased excitability: they were in a state of constant motor excitation, they reacted extremely violently to the approach of their "master," and so on.

Pedometer recordings (during eleven night hours) of the motor activity of one of the hybrids and of a Husky demonstrated a clear difference between them (Table 11). (The pedometers were attached to the dogs for five days; the dogs were kept in neighboring cages so that the external stimuli which could induce a state of excitement in the dogs were identical.)

Table 11. Recordings of the Motor Activity by Means of a Pedometer

Name	Breed	1st day	2nd day	3rd day	4th day	5th day	Mean
Chavrkin	Husky.	2,929	1904	3100	2,534	1,328	2,359
Zhenya	Hybrid of Husky and Shepherd	13,500	8934	7693	15,376	14,500	12,000

The results in Table 11 illustrate the difference between the motor activity of the hybrid and the Husky (the former made five times more movements than the latter).* This confirmed the first point of our working hypothesis that the greater excitability of hybrids, as compared to Huskies, may possibly be the cause of the manifestation of their passive defensive reaction.

To verify the hypothesis that Huskies possess an unconditioned reflex basis for manifestation of a passive defensive reaction which they do not in fact display because of their low excitability, we conducted experiments in which their excitability was increased. To cause excitation, cocaine in a dose of 2.5 mg/kg body weight was injected subcutaneously into the dog. A change in the dog's behavior usually began 20-30 minutes after the injection and lasted about 1.0-1.5 hours.

As a preliminary, to test the exciting action of cocaine, dogs of high and average levels of excitation and not possessing passive defensive reactions were selected. Cocaine was injected into 12 dogs. Their motor activity in the course of one hour before and one hour after injection was compared. The experiments showed that cocaine, in the dose used, considerably increased the excitability of the experimental dogs (the mean pedometer reading rose

---

*Similar findings characterizing all hybrids (which were more excitable than Huskies) are cited in our paper "Investigation into the phenogenetics of behavioral characteristics in dogs." Biologicheskii Zhurnal, 7, 4 (1938).

from 480 to 2800). No passive defensive reaction was displayed in any of the animals.

After these preliminary experiments we proceeded with experiments with Huskies. Unfortunately, at this time only one of the three Huskies used for hybridization remained alive—the male dog Chavrkin. In the nursery kennels there was another female Husky, Changush, brought from the same district near the Amur River, and her offspring, a male dog Ingush. In their behavior they were equally unexcitable and displayed no passive defensive reaction whatever. After injection of cocaine, however, besides increased general excitability, all three dogs showed a well-marked passive defensive reaction. When the action of cocaine ceased, this reaction disappeared, parallel with a decrease in the state of excitation of the dog. We give below an extract from the records of an experiment.

When I approached the cage, the dog (the male Chavrkin) sat by the bars. As I came up to the bars the dog continued to sit, then slowly stood up, went away from the bars and lay in the shade. I knocked on the bars several times with my hand; the dog stood up, crossed to another place and lay down again. I went inside the cage and the dog continued to lie in the same place. I went up to the dog, which let itself be stroked.

14 hr, injection of 62.5 mg cocaine.

14 hr 45 min. When I approached the cage the dog sat by the bars. As I came up to the bars the dog lay down, gave a start at the noise, raised its ears but continued to lie. I went inside the cage; the dog stood up and, with its tail down, went away to the end of the cage, where it sat for a little while.

15 hr. When I approached the cage Chavrkin went to the other end. I went inside the cage and called the dog. It turned in my direction, but stopped after two or three paces, crouched down, and pressed back its ears. As I approached the dog it ran away to the end of the cage. I went up to the dog and it crouched silently. Next day Chavrkin no longer showed the passive defensive reaction.

The manifestation of a passive defensive reaction after the injection of cocaine into these three dogs (and especially in Chavrkin) conformed fully to the anticipated result. The injection of cocaine led to increased excitability and enabled the display of a genotypic passive defensive reaction, the threshold of which was so high in the normal state of the dog that this particular reaction did not appear.

Further investigations showed that not only the manifestation itself, but also the degree of expression of an already existing passive defensive reaction in a dog is dependent upon the excitability of the animal. Injection of cocaine into seven Huskies whose behavior showed a passive defensive reaction, resulted in increase of this reaction in six of these animals, and an unchanged degree of reaction in only one. Increased excitability caused by injection of cocaine thus not only leads to the manifestation of a passive defensive reaction, but also increases the degree of its expression.

In Pavlov's school it was considered that the presence of a
passive defensive reaction is characteristic of a weak type of
nervous system. Rozental' (1936) pointed out that dogs with this
type of nervous system cannot overcome the "childish" reflex of
timidity, and they remain cowardly throughout life. In dogs with
a weak type of nervous system, caffeine in certain doses does not
increase excitability, as is observed in dogs of a strong type, but
on the contrary lowers it, "falling outside the limits of the working
capacity of the cell" (Pavlov, 1935).

It might be postulated from these findings that cocaine, like
caffeine, causes inhibition rather than excitation in dogs with a
weak type of nervous activity (but in which a passive defensive
reaction is displayed). This takes the form of a strengthening of
the passive defensive reaction. In other words, it may be that it is
not an increase in the excitability of the nervous system which leads
to the strengthening of the passive defensive reaction but, on the
contrary, a limiting inhibition of the nervous system that develops
under these circumstances, and is expressed as a strengthening
of the passive defensive reaction.

In order to elucidate this problem, we conducted experiments
to determine the state of excitability of dogs with a passive
defensive reaction after these animals were given injections of
cocaine. The experiments were conducted as follows. The defensive
behavior of the dog was first described. A pedometer was then
attached for one hour, after which an injection of cocaine was given.
From 20 to 30 minutes later the changes in the dog's behavior were
noted, and the pedometer was again attached for one hour. Comparison of the pedometer readings before and after the injection
showed the changes in the excitability of the dogs.*

Before carrying out the main experiment it was necessary to
find out how taking the dog to the laboratory for the injection and
the injection itself were reflected in the defensive behavior and the
general excitability of the animal. For this purpose experiments
were carried out in which dogs were injected with physiological
saline, and these experiments showed that the procedure did not
affect the behavior of the dogs; the pedometer readings also were
practically unchanged.

Injection of cocaine into 20 dogs possessing a passive defensive
reaction led in 18 animals to a strengthening of this reaction; in

*The dogs were kept before and after injection in the cages in which they lived.

16 the motor activity was also increased. In only four of the 20 dogs was the motor activity diminished after the injection of cocaine.

These investigations thus yielded the following results. Although cocaine, in the dose injected, leads to an intensification of the passive defensive reaction, at the same time it increases the general excitability of the nervous system of the dogs. This suggests that the strengthening of the passive defensive reaction after the injection of cocaine is not, in fact, the result of the development of limiting inhibition of the nervous system, but is the result of an increase in the degree of its excitability.

When summing up our experiments on the relationship between the manifestation and expression of the passive defensive reaction and the degree of general excitability, it must be emphasized that researches using absolutely different methods (by means of genetic hybrid analysis or the use of drugs which increase excitability) gave identical results. On the basis of our findings it can be asserted that the manifestation and expression of a passive defensive reaction is dependent upon the general excitability of the animal: the higher the general excitability, the stronger the degree of expression of this reaction; and conversely, with a low degree of excitability, the innate passive defensive reaction may absolutely fail to manifest itself in the dog's behavior.

Our findings afford further confirmation of the correctness of the views expressed by Pavlov during the last years of his life (he would not agree to regard cowardice in dogs as the opposite of excitation). These experiments not only indicate that cowardice cannot be opposed to excitability, but they also show that an increase in excitability leads to a strengthening of the passive defensive reaction.

We investigated the importance of the degree of excitability of the nervous system for the expression of the active defensive reaction in experiments similar to those used to investigate the passive defensive reaction.

Nine dogs, possessing an active defensive reaction, were given injections of cocaine (in a dose of 2.5 mg/kg body weight), and the changes in the expression of the defensive reaction were subsequently recorded. It was found that in all the dogs the increase in excitability arising as a result of the injection of cocaine led to an intensification of the expression of the active defensive reaction.

The intensification of the active defensive reaction after injection of cocaine is also clearly seen from a comparison of the kymograms of the barking of the dogs in response to a stimulus causing this reaction, before and after injection (Fig. 9).

The results obtained thus show that the expression of this unitary reaction of behavior also bears a positive relationship to the excitability of the nervous system.

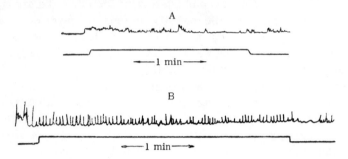

Fig. 9. The active defensive reaction before (A) and after (B) injection of cocaine. (Recording of the barking of a dog.)

Let us examine the relationship between the active and passive defensive reactions with regard to the general level of excitability. In Chapter 1 it was shown that it is possible to dissociate the defensive form of behavior of "aggressive-cowardly" dogs into individual unitary reactions: active and passive defensive. In the preceding sections of this chapter it has been shown that the degree of expression of each of these reactions is dependent upon the general level of excitability of the dog. The question arises whether the mutual relationship of the individual unitary reactions which are combined to construct a particular biological form of behavior in animals is dependent on the general excitability.

We undertook the study of the mutual relationship of the active and passive defensive reactions. Experiments were carried out on aggressive-cowardly dogs in which the excitability was increased by the injection of cocaine (2.5 mg/kg body weight), and the changes in the expression of their defensive reactions of behavior were studied (Table 12).

It can be seen from Table 12 that the increase in excitability arising as a result of the injection of cocaine led to different effects in different dogs. In one group of dogs (Nos. 1-7) the expression of both active and passive defensive reactions was strengthened, and in another group a strengthening of the passive defensive reaction was accompanied either by the weakening of the expression of the active defensive reaction (Nos. 10 and 11), or by its complete

Table 12. Results of Injection of Cocaine into Aggressive-Cowardly
Dogs

No.	Name of dog	Passive defensive reaction	Active defensive reaction
1	Rustan	Strengthening	Strengthening
2	Nora	Strengthening	Strengthening
3	Narfa	Strengthening	Strengthening
4	Dzhil'da	Strengthening	Strengthening
5	Reks	Strengthening	Strengthening
6	Krak	Strengthening	Strengthening
7	Dzhek	Strengthening	Strengthening
8	Dina	Strengthening	Insignificant strengthening
9	Allyur	Strengthening	Unchanged
10	Nelli I	Strengthening	Weakening
11	Efrat	Strengthening	Weakening
12	Zhar	Strengthening	Completely absent
13	Kom	Strengthening	Completely absent
14	Gerta	Strengthening	Completely absent
15	Askhari	Weakening	Strengthening
16	Til'da	Weakening	Strengthening
17	Taiko	Weakening	Strengthening
18	Grei	Almost complete disappearance	Strengthening
19	Al'ma	Unchanged	Unchanged
20	Nelli II	Unchanged	Unchanged

disappearance (Nos. 12, 13, and 14). In a third group of dogs the converse phenomenon was observed: a strengthening of the expression of the active defensive reaction was accompanied by a weakening of the expression of the passive defensive reaction, and even by its complete disappearance (Nos. 15, 16, 17, and 18).

These experimental results showed that when the level of excitability of the nervous system is increased, dogs in which the two unitary reactions are manifested simultaneously behave differently from dogs in which only one of these reactions is displayed in the animal's behavior. In the presence of only one unitary reaction this is strengthened. If two unitary reactions are displayed in the animal's behavior, then not only may both be simultaneously strengthened, but sometimes one may be strengthened while the other is inhibited. We regard this latter phenomenon as the result of the inductive inhibition of one reaction by the other.

A case of the inhibition of one reaction by another is also described in the paper by Pavlov and Petrova (1916). By exciting (strengthening) first the "guarding reflex" (active defensive), then the food reflex, they were able to observe the suppression of one reaction by another. "The two reflexes literally behave like two scale pans. You have only to increase the number of stimuli for one reflex, i.e., as it were to add a small weight to one pan, for it to begin to tip the scales; this reflex suppresses the other, and vice versa. If stimuli are added to the other, we now see how this becomes dominant, i.e., the corresponding pan now tips the scales."*

Although our case and that described by Pavlov and Petrova are phenomena of the same order, there is a difference between them. Pavlov and Petrova increased one of the reactions and observed inhibition of the other. After injection of cocaine, as a result of the increase in the general excitability, both reactions were strengthened. However, since the active and passive defensive reactions are diametrically opposite in their direction (one toward flight, the other toward fight), the increase in the degree of their expression must lead to the strengthening of the negative induction relationships existing between them.

In those cases when one defensive reaction in aggressive-cowardly dogs was weaker than the other, an increase in the degree of expression of both (after the injection of cocaine) was also manifested as a relative increase in the strength of one reaction (the stronger). This led to inhibition of the weaker reaction. In those cases when both reactions were equal in strength, an increase in the level of excitability, strengthening the expression of both reactions, did not disturb their equilibrium.

Experiments in which cocaine was injected into aggressive-cowardly dogs thus clearly indicated the existence of negative induction relationships between the active and passive defensive reactions, and the possibility of the total inhibition of one defensive reaction by the other as a result of an increase in general excitability.

The possibility of complete inhibition of the passive defensive reaction by the active is demonstrated convincingly by a series of cases. We carried out experiments on a dog with the name Dzherri in whose behavior both components of the defensive behavior were well marked. We give below a description of the defensive behavior of this dog.

Dzherri is a German shepherd, male, aged 3 years.
July 15. "When I approached the cage in which Dzherri was sitting, the dog stood up by the bars and barked lethargically. When I was at a distance of five paces from the cage, the dog retreated one or two paces from the bars, but then came back barking. When I came still nearer (two or three paces from the cage) the dog stood by the bars and barked, and in time ran away into the hut (a covered building), and then again came out barking to the bars. I came right up to the bars; the dog ran away into the hut and barked from there; on one occasion it bounded out barking, but without coming as far as the bars it ran back again into the hut. After I had stood quietly and motionless by the

*I. P. Pavlov, op. cit., p. 323.

bars for two minutes the barking almost ceased, although from time to time the dog gave a slight whimper and surveyed the scene from the hut. I went away from the cage, and the dog immediately bounded after me with an aggressive bark and a growl. When I entered the cage in which the dog was sitting, Dzherri ran away and hid in the corner of the hut; when I called the dog it did not come out, but stood sulkily in the corner of the hut with its tail down and its ears pressed back. When I left the cage and started to walk away, the dog bounded after me with a bark."

We then began to develop (by training) the active defensive component of Dzherri's behavior. The dog's behavior changed greatly. All traces of the passive defensive reaction disappeared completely and Dzherri became one of the most aggressive dogs in the kennels.

November 8. "When I approached the cage, Dzherri bounded to the bars with a bark. As I drew nearer the bark became louder and the dog turned towards the bars and jumped up at them. As I came up to the cage the dog barked and jumped up at the bars. I stood quietly by the cage and the dog barked furiously. After 3 min 30 sec there was a short interlude in the barking (of 2-3 sec), and after 4 min 15 sec a longer interlude in the barking, of 10 sec. I stood there for 6 min trying not to make any movements; the dog barked, but at intervals of 10-12 sec. When I made the slightest movement the barking became stronger and the dog sprang up the bars, trying to grab me. When I opened the cage door a very little, the dog hurled itself at the door, so that it was immediately closed with a slam. As I went away from the cage, Dzherri continued to bark and paced up and down by the bars."

It is clear from the description given that the dog now showed complete absence of a passive defensive reaction, and displayed a well-marked aggressiveness. The question arises, what had happened to the passive defensive reaction—why had it disappeared? It is perfectly natural to suggest that the strong expression of the active defensive reaction had completely inhibited the passive. If this is the correct explanation, then the weakening of the expression of the active defensive reaction would be bound to lead to the manifestation of a passive defensive reaction, and it would therefore be expected that by giving the dog an injection of morphine, which weakens the active defensive component of the defensive form of behavior, the manifestation of the inhibited passive defensive reaction would result.

February 11. "12 hr 55 min—a pedometer was attached to the dog. 1 hr 55 min—pedometer removed, reading 650. 2 hr 30 min—injection of morphine (0.03 g). 3 hr 15 min—pedometer attached to the dog. 4 hr 15 min—pedometer removed, reading 300. 4 hr 20 min—when I approached the cage Dzherri was standing by the bars. I was at a distance of 20-30 paces from the dog it barked, although lethargically. I came up to the cage; the dog barked a few times and went away into the hut with its tail down. It stood for a few seconds, looking around, and then went inside. I knocked on the bars with my hand; Dzherri bounded out of the hut with a bark, but before reaching the bars near which I was standing, ran back into the hut. As I went away from the cage, the dog bounded out of the hut with a bark. As I walked away the dog again jumped after me with a bark."

Next day Dzherri's behavior was just as it had been before the injection of morphine: the dog began to exhibit the active defensive reaction only. The description of Dzherri's behavior after the injection of morphine strikingly resembles that observed in the dog before the active defensive reaction had begun to develop.

Let us consider the opposite case—the inhibition of a well-marked passive defensive reaction by the active defensive component of behavior. We investigated such a case in hybrids of Huskies and shepherds (see pp. 65-66). A small proportion of the hybrids exhibited, besides a passive defensive reaction, an active defensive reaction which they inherited from their shepherd mothers. The hybrids which displayed an active defensive reaction possessed a less marked passive defensive reaction. An increase in the excitability of these dogs as a result of the injection of cocaine led to a strengthening of the passive defensive reaction and to a weakening of the active defensive reaction (Table 12, Efrat) or even to the total inhibition of its manifestation (Table 12, Zhar). This led us to believe that hybrids with a well-marked passive defensive reaction display no active defensive reaction in their behavior because it is inhibited by the passive defensive reaction.

In order to verify this hypothesis, the hybrid Zhenya, possessing a well-marked passive defensive reaction but not displaying an active, was crossed with a Zyryan Husky, Sos, that demonstrated no defensive reactions. A litter of eight puppies was obtained from this crossing. If we assume that an active defensive reaction did not show itself in Zhenya's behavior because the passive defensive reaction was well marked, it might be expected that the dogs born in this litter (as a result of dissociation) would be aggressive, aggressive-cowardly, cowardly, or devoid of defensive reactions.*

It can be seen from Fig. 8 that among the offspring of Zhanya and Sos, reared in the same conditions, all the possible combinations of active and passive defensive behavior were found.

The appearance in the litter of individuals with active defensive reactions confirmed our hypothesis that this reaction did not appear in many of the hybrids of Huskies and German shepherds (including Zhenya) because they had a well-marked passive defensive reaction.

---

*If the passive defensive reaction is denoted by T, its absence by t, the active defensive reaction by A and its absence by a, we may represent this particular crossing and its result by the following scheme:

$$\text{Zhenya} \times \text{Sos}$$
$$\text{TtAa} \quad \text{ttaa}$$

TtAa	Ttaa	ttAa	ttaa
Agressive-cowardly or cowardly	Cowardly	Aggressive	Absence of defensive behavior

Undoubtedly the offspring of Zhenya and Sos inherited an active defensive reaction from Zhenya (whose mother was aggressive), for neither Sos nor any of his offspring obtained from other females showed even traces of this reaction.

One of the aggressive-cowardly offspring of Zhenya and Sos was given an injection of cocaine which, leading to an increase in excitability, completely inhibited its active defensive reaction. The passive defensive reaction was accordingly considerably increased in strength. The dog became as cowardly as its mother.

The parallel investigations of this litter by a genetic method and by the use of drugs thus gave identical results. As a result of negative induction relationships, a well-marked passive defensive reaction may completely inhibit the manifestation of an active defensive reaction.

From the foregoing remarks the following conclusion may be drawn. When two unitary reactions, for example active and passive defensive reactions, are manifested simultaneously in an animal's behavior, an increase in the excitability of the nervous system, leading to an increase in the strength of each of them, will also increase the negative induction relationships existing between them (if such do, in fact, exist), which may lead to the total inhibition of one reaction (the "weaker") by the other (the "stronger").

The findings described illustrate the important relationship existing between the manifestation and expression of behavioral acts of animals and the functional state of their nervous system.

The manifestation and the expression of different behavioral acts are largely dependent upon the general state of excitability of the nervous system. Against a background of increased excitability of the nervous system, the various behavioral acts of animals are most sharply expressed. Conversely, against the background of low excitability of the nervous system, behavioral acts the manifestation of which is determined genotypically or by suitable conditions of training may generally be absent from the animal's behavior.

Individual differences in the degree of excitability of the nervous system may thus be responsible for considerable individual differences in the behavior of animals. These differences are displayed in behavioral acts the formation of which takes place under the dominating influence of individual experience or of innate components.

Let us examine the formation of behavior as it depends on the strength of the animal's nervous system. Pavlov (1935) suggested special methods of determining the strength of the nervous system.

All these methods were directed at detecting the limit of working capacity of the nervous system by increasing the tension of the processes of stimulation and inhibition. "The importance of the strength of nervous processes is clear from the fact that unusual, excessive events and stimuli of great strength are present more or less frequently in the external environment, so that the necessity often may arise of suppressing or restraining the effects of these stimuli to meet the requirements of other similar, or still more powerful external conditions."[*]

In animals with a weak nervous system a very slight increase in the tension of the two processes leads to disturbance of the normal conditioned reflex activity. In animals with a strong nervous system, even when the two processes are under considerable strain, not only is the conditioned reflex activity not disturbed but, on the contrary, it is improved. In addition to these extreme cases there is a continuous series of quantitative gradations. Thus, by strength of the nervous system we must understand (according to Pavlov's view) its ability to withstand considerable strain of the processes of stimulation and inhibition, and to maintain its normal working capacity under these circumstances.

At Pavlov's initiative, work to study the genetics of the higher nervous activity of dogs was started in the early 1930s at the biological station in Koltushy. Pavlov himself was unable to see these investigations to their conclusion. After his death the work was continued by the staff of the Pavlov Institute of Physiology.

Kolesnikov (1947) studied a line of dogs characterized by considerable weakness of the nervous system. All the dogs of this line (descended from a female with a very weak nervous system) possessed weakness of the nervous system transmitted from generation to generation. Krasuskii (1953) concluded from the resemblance between four offspring and their mother, characterized by inertia and imbalance of the nervous processes, that these properties of the nervous system were inherited in dogs.

We (Krushinskii, 1947a) analyzed data concerning the inheritance of the strength of the nervous system in dogs from the kennels of the Pavlov Institute of Physiology in Koltushy. The results obtained are given in Table 13.

The results of the crossings, and also the coefficient of correlation between the strength of the nervous system in the siblings, with values of $+0.34 \pm 0.10$ in individual litters, illustrate that

[*]I. P. Pavlov, op. cit., p. 652.

Table 13. Inheritance of the Strength of the Nervous System in Dogs

Strength of nervous system in parents	No. of litters	Number of offspring	
		of strong type	of weak type
Strong x strong	6	16	4
Strong x weak	3	7	5
Weak x weak	1	0	6

differences in the strength of the nervous system in dogs are determined by the genotype. The impression may even be created
that a weak type of nervous system is inherited as a recessive
sign. It is hardly possible, however, to deduce such a conclusion
(because of the insufficiency of the data). At the same time the
results undoubtedly show that the strength of the nervous system
is a property the formation of which, as Pavlov supposed, is largely
determined by the genotype of the animal.

The investigations of foreign workers have yielded similar
results. Humphrey and Warner (1934) showed that selection improved the trainability of a population of dogs. Dawson and Katz
(1940) found that there is a great difference in the speed of formation of reflexes both between individual dogs and between breeds.
Deliberate selection increased the mean values of the speed of
formation of conditioned reflexes in the offspring by comparison
with the mean values for the population as a whole.

Pawlowski and Scott (1956) found a considerable difference
between various breeds of dogs in the degree of "submission" of
some individuals to others when they were kept together, which
was explained by these workers as due to genotypic differences in
their nervous activity. There is no doubt that the strength of the
nervous system must play a fundamental role in the phenomenon
of "dominance of behavior" of some individuals over others.

Investigations on other animals confirmed the role of heredity
in determining the individual modalities of nervous activity. Yerkes
(1916) studied the ability of two lines of rats to undergo training:
one noninbred, the other developed by close inbreeding over a long
period. The rats were trained in a circular maze and in a Yerkes
differential chamber. It was found that the noninbred rats learned
more quickly than the inbred.

Bagg (1916) studied the speed of training of two lines of mice
in a simple maze. Lines of white and yellow mice were crossed.

It was found that the average learning time of the line of white mice during the last 15 experiments was 27.5 ± 2 sec, with nine mistakes per experiment; the yellow— 83.0 ± 7.0 sec, with two mistakes per experiment. Both of these investigations indicate the role of inherited factors in the ability of mice to learn in a maze.

Tolman (1924) showed that the average ability of rats to form conditioned reflexes may be modified by means of selection. Vicari (1929) studied the role of heredity in the speed of formation of conditioned reflexes during training of mice of various lines in a maze. He found that individual lines of mice have their own characteristic curves of the course of formation of the conditioned reflex of passing through a maze. Crossing lines of mice slow and quick to form conditioned reflexes showed that the first generation hybrids closely resemble in this property the lines of mice quick to form conditioned reflexes. In the second generation dissociation was observed. The large number of cases (900) investigated and the detailed analysis of the results obtained, indicate the role of heredity in the speed of formation in mice of the conditioned reflex of passing through the maze. These researches thus showed that individual variations in the general properties of the nervous system are largely determined by the genotype of the animal.

Pavlov clearly described the behavior of dogs with different types of nervous activity. Speaking of dogs of weak type, he stated: "We must recognize a type of weak animals, characterized by obvious weakness of both stimulation and inhibition, never fully adapted to life and easily overwhelmed, quickly and often becoming ill or neurotic under the influence of difficult situations in life or, what amounts to the same thing, during our difficult nervous experiments. The most important factor of all is that this type, as a rule, cannot be greatly improved by training and discipline, and can successfully adapt only in certain particularly favorable, deliberately created conditions or, as we usually call it, in 'hothouse' conditions."* The importance of the strength of the nervous system in the animal's life is clearly seen from this quotation.

As an illustration of Pavlov's words cited above we may present our findings on the connection between the quality of the response of dogs to training in certain motor skills and the strength of their nervous system.

In Pavlov's school, in order to determine the limits of working capacity of the nerve cells a single indicator was used—the con-

*I. P. Pavlov, op. cit., p. 675.

ditioned reflex activity. However, the unconditioned reflex activity may also be used to determine this general property of the nervous system.

Although at present the original standard for determination of the strength of the nervous processes has been modified (Kolesnikov and Troshikhin, 1951; Maiorov and Troshikhin, 1952; Kolesnikov, 1958), the basic principle of the determination has remained the same as that formulated by Pavlov—establishment of the limit of working capacity of the nervous system by means of its excitation by the action of strong stimuli.

In order to determine the strength of the nervous system we employed strong sound stimuli. The ability of the animal to withstand sounds of different strengths, on the one hand, and the speed with which a passive defensive reaction in relation to this stimulus is inhibited, on the other hand, were used as indices of the strength of the nervous system. As stimuli we used the sound of a rattle or of an automobile horn, applied when the dogs were eating their food (Krushinskii, 1946).

According to their reactions to these stimuli, the dogs were divided into four groups. 1) Dogs not withstanding either stimulus; we defined these as animals with a weak nervous system. 2) Dogs withstanding both stimuli with difficulty (approaching their food after successive withdrawal and reapplication of the stimulus). We classed such animals as the strong variant of the weak type. 3) Dogs withstanding one stimulus or the other with difficulty. This was the weak variant of the strong type. 4) Dogs withstanding both stimuli. We classed these as the strong type.

The coefficient of correlation between the reactions of the dogs to the two stimuli which we used had the value of $+0.58 \pm 0.09$.

We fully realized that it is impossible to give a complete description of the strength of a dog's nervous system on the basis of the animal's reaction to strong sound stimuli. There is no doubt, however, that differences in the reactions of dogs to standard sound stimuli are mainly due to the typological characteristics of the nervous system. Dogs with a weak nervous system are less able to withstand strong sound stimuli than are those with a strong nervous system.

The study of the relationship between the strength of the nervous system and the success of production of individually acquired habits was undertaken on four groups of dogs trained for the antitank and communications services during wartime. The conditions in which the dogs were trained differed.

The greatest effort during training was required from the dogs of group 2; they were trained in antitank work in the difficult time of July-August 1941, when it was necessary to mobilize urgently all possible antitank resources. Training of this group of dogs was carried out twice a day. The dogs were intensively taught to accustom themselves to the action of excessively strong sound stimuli. These procedures which were to form the conditioned reflex of penetrating between the caterpillars of a moving tank, in the immediate neighborhood of which excessive strong sound stimuli were operating, undoubtedly put a very great strain on the nervous system of the dogs. Groups 1 and 4 were trained in a somewhat later period, when the urgency of the work was not so great. Acclimatization to battle sounds was also undertaken less intensively. Finally, group 3 was trained at a still later period of the war. The duration of training of the dogs was lengthened and their acclimatization to gunfire was carried out gradually. According to degree of difficulty, the conditions in which the dogs were trained could therefore be arranged in the following order: group 2 dogs were trained in the most difficult conditions, group 3 dogs in the easiest conditions, and groups 1 and 4 in conditions of medium difficulty.

The coefficients of correlation between the degree of strength of the nervous system and the quality of the dogs' work during training are shown for each group in Table 14.

The results described show that a slight correlation is observed between the strength of the nervous system and the formation of these particular individually acquired skills only in relatively difficult conditions of training of the dog. In easy training conditions no correlation could be found.

Jointly with Fless, we made a further study of the relationship between the formation of individually acquired skills and the strength of the nervous system in trailer dogs. The strength of the nervous system was determined in two ways: 1) by the action of excessively strong stimuli (a rattle and a siren), and 2) by the administration of large doses of caffeine.

The latter method was as follows. By means of a pedometer we determined the degree of "food" and "active defensive" excitability. These tests were then repeated after administration of different doses of caffeine (0.3 and 0.8 g). The lowering of the indices of the degree of motor excitability indicated that the limits of working capacity of the nerve cells had been exceeded. The magnitude of the dose of caffeine at which the excitability was still not lowered characterized the strength of the dog's nervous system.

Table 14. Coefficients of Correlation between the Quality of Formation of an Individually Acquired Skill and the Strength of the Nervous System in Dogs

Group No.	Conditions of training	n	Coefficient of correlation $r \pm m_r$	Reliability of correlation r/m_r
3 (AT1)	Easy	54	+0.06 ±0.14	0.43
1 (Communications)	Medium	54	+0.27 ±0.13	2.08
4 (AT2)	Medium	61	+0.29 ±0.12	2.42
2 (AT1)	Difficult	100	+0.33 ±0.09	3.67

The results obtained were compared with the indices of the response to training both by means of "general training"* and by the special methods of training used in the trailer dog service—work in trailing human beings. The qualities of the training response were compared in dogs "withstanding" and "not withstanding" the action of strong stimuli when "general training" and trailing work were used.

The results were also compared with the indices of the strength of the nervous system obtained as the result of tests of the action of caffeine.

The results obtained suggest a positive relationship between success in the formation of these particular individually acquired skills and the strength of the nervous system. Under these circumstances the formation of the more difficult skill (trailing work) bore a closer relationship to the strength of the nervous system than did the formation of the less difficult skills of "general training" (Fless, 1952).

From the findings described above, the following conclusion may be drawn. Whereas in easy conditions the formation of individually acquired motor skills is almost independent of the strength of the nervous system, as the difficulty of the conditions or the complexity of the skill itself increases, the strength of the nervous system acquires ever greater importance to the success of formation of skills.

Let us examine the relationship between the unitary reactions of behavior (passive and active defensive) and the strength of the nervous system.

*In general training a series of conditioned reflexes were formed in the dogs: approaching its master when summoned, sitting down, lying down, etc.

The study of the passive defensive reaction was undertaken in Pavlov's school in connection with the study of the types of nervous activity in dogs. The passive defensive reaction was first described by Frolov (1925). It was subsequently investigated by several workers (Rikman, 1928; Speranskii, 1926, 1928; Vyrzhikovskii, 1928). They regarded this reaction as a manifestation of the weakness or state of inhibition of the cortical cells. Pavlov (1927) considered that "a simple predominance of the physiological process of inhibition over the expression of weakness of the cortical cells lies at the basis of normal fear, cowardice, and in particular the pain phobias."

Subsequently, however, Ivanov-Smolenskii (1932), who carried out a special investigation on a cowardly dog, showed that in spite of the presence of a well-marked passive defensive reaction in this dog, the animal could not be classed as having a weak type of nervous system. This research showed for the first time that cowardly dogs do not necessarily possess a weak nervous system.

Training of puppies in isolated conditions (Vyrzhikovskii and Maiorov, 1933; Vyrzhikovskii, 1937) showed that such conditions of training are responsible for the manifestation in a dog of a passive defensive reaction which is preserved for a long time in the individual life of the animal (Zeval'd, 1938).

Rozental' (1936) carried out special experiments on a dog with a passive defensive reaction which had for several years been classed as a weak type, and showed that this dog should in fact be classed as a strong type. This worker suggested that the dog had not "shaken off the infantile fear reflex." While he did not consider that the passive defensive reaction must inevitably be associated with a weak type of nervous system, Rozental' believed that puppies of a strong type with age may shed their "infantile" reaction of caution, whereas in puppies of weak type this reaction remains to some extent throughout life, as it does in dogs of the strong nervous type but brought up under particularly unfavorable conditions.

On the basis of these facts, Pavlov in the last years of his life dissented from the original identification of cowardice and weakness of the nervous system: "All apparently cowardly dogs, i.e., dogs slow to accustom themselves to the circumstances of our experiments, which also had difficulty in forming conditioned reflexes (and the whole of their conditioned reflex activity was easily disturbed by new, external influences of no great significance), have been classed by us indiscriminately as having a weak type of nervous system."* Pavlov thus considered cowardice as behavior which constantly masks the true strength of the nervous system.

*I. P. Pavlov, op. cit., p. 655.

The problem of the relationship between the formation of a passive defensive reaction and the strength of the nervous system was investigated by us in two groups of dogs.

1. Mongrel dogs (51 individuals), nearly all of which had been reared in the nursery kennels of the Pavlov Institute of Physiology, i.e., in conditions of considerable isolation. The strength of the nervous system in these dogs was determined by a method devised while Pavlov was still alive.* This method is based on four indices (Timofeev, 1941): 1) the speed of formation of the first positive conditioned reflex; 2) the influence of fasting for one day on the magnitude of the conditioned reflex; 3) the influence of various doses of caffeine on the conditioned reflex activity; 4) tests of the conditioned reflex activity of.the dog with excessively strong stimulation (with a rattle).

2. Dogs (229 individuals) belonging to various breeding kennels. The very great majority of these animals were German shepherds. The strength of the nervous system of the dogs was determined by their reaction to strong sound stimuli, and it was compared with the expression of their passive defensive reaction. All the dogs had been brought to the kennels by private individuals, i.e., they had been reared in conditions of relative freedom.

The coefficient of correlation between the strength of the nervous system and the absence of a passive defensive reaction in the dogs of the Koltushy kennels was +0.34 ±0.22.† In Table 15 we present the results showing the frequency of the combination of a passive defensive reaction with weak and strong types of nervous system.

It will be seen from Table 15 that a passive defensive reaction is freely combined with both strong and weak types of nervous system, although the overwhelming majority of dogs without a passive defensive reaction were of the strong type.

*The types of nervous system of the dogs were determined by the following members of the staff of the Institute: T. A. Timofeeva, V. A. Troshikhin, Z. V. Troshikhina, L. O. Zeval'd, M. S. Kolesnikov, V. K. Krasuskii, K. N. Bolokhovskii, V. F. Pleshkov, and others.

†Since the dogs of this population belonged to several families, the method of calculation of the coefficient of correlation differed from that used for the freely crossed population. Calculation of the error was undertaken by the formula proposed by M. V. Ignat'ev, to whom I wish to express my appreciation.

$$m_r = \frac{1 - r^2}{\sqrt{\dfrac{n}{K}}} \; ;$$

the value of K was calculated from the formula:

$$K = 1 + {}^1/_4 \left[-\frac{n'(n'-1) + n''(n''-1) + \ldots\ldots -2L}{N} \right],$$

where n' is the total number of individuals in the first family; n″ the number in the second family, and so on; L is the number of families, and N the number of individuals studied.

Table 15. Coexistence of a Passive Defensive
Reaction with Weak and Strong Types of Nervous
System in Dogs

Defensive reactions	Of weak type	Of strong type
With a passive defensive reaction	16	19
Without a passive defensive reaction . .	1	15

A similar pattern of relationship between the passive defensive reaction and the strength of the nervous system is also observed among the dogs of the second group. The coefficient of correlation between the strong nervous system and the absence of a passive defensive reaction has a value of $+0.33 \pm 0.06$.

In Table 16 we present data illustrating the frequency with which a passive defensive reaction is found in dogs which could withstand and those which could not withstand a strong sound stimulus.

Table 16. Coexistence of a Passive Defensive Reaction with Variation in the Reaction of the Dogs to a Sound Stimulus

Group of dogs	Not withstanding sound stimulation (weak nervous system)	Withstanding sound stimulation (strong nervous system)
With a passive defensive reaction	25	34
Without a passive defensive reaction	21	149

The figures in Table 16 demonstrate that a passive defensive reaction in the dogs of this group may be freely combined with a weak and a strong nervous system. However, in the overwhelming majority of cases, the dogs which did not display cowardice possessed a strong nervous system.

The findings described thus showed the presence in the two groups of dogs of a similar relationship between the passive defensive reaction and the strong nervous system. This indicates that the relationship which we have been studying is based on certain definite rules. We consider that two conclusions may be drawn.

First, the free combination of a passive defensive reaction with both a strong and a weak type of nervous system suggests that the strength or weakness of the nervous system are not themselves the factors responsible for the formation of the passive defensive reaction.

Second, the considerable predominance of dogs without a passive defensive reaction among individuals with a strong nervous system indicates that the strong nervous system is a favorable condition for the formation of behavior in which a passive defensive reaction is absent.

It is evident that a weak nervous system is not in itself the cause of development of a passive defensive reaction. The development of the latter requires specific factors: the presence of an innate predisposition, or of unfavorable conditions of rearing. However, dogs possessing a strong nervous system, can more easily "get rid of" or inhibit a passive defensive reaction in the course of life. With unfavorable (isolated) conditions of rearing, when it is difficult for the puppy to shed the passive defensive reaction, the presence of a strong nervous system is particularly important to the dog's ability to lose its cowardice.

In fact, as is clear from Table 15, among the uncowardly dogs of the Koltushy nursery (reared in conditions of partial isolation) the dogs with a strong type of nervous system predominate over those with a weak type, in a ratio of 15 : 1.

To summarize the foregoing remarks, we may say that our findings demonstrate not so much the presence of a connection between a weak nervous system and a passive defensive reaction as the presence of a relationship between a strong nervous system and the absence of a passive defensive reaction. The strength of the nervous system is a condition enabling the passive defensive reaction to be inhibited in the period of formation of the dog's behavior.

The active defensive reaction has been studied less than the passive by Pavlov's school. Pavlov (1927) pointed out that dogs of an excitable type of nervous system (strong, unbalanced) are aggressive. After this remark had been made, the active defensive reaction was considered by Pavlov's school to be an indication of the strength of the nervous system. For example, in the book edited by Pavlov's pupil Andreev (1939), the active defensive reaction was associated with the strong, excitable type of nervous system. Yakovleva (1940), describing the active defensive behavior of her experimental dog, came to the following conclusion: "Such aggressive behavior in a dog convinced us that its nervous system was of the strong type."

Thus although there is no concrete experimental research to shed light on the relationship between the strong type of nervous system and the active defensive reaction, nevertheless views have been expressed that such a relationship exists.

In our research, as material for the establishment of a relationship between the active defensive reaction and the strength of the nervous activity (determined by the reaction of the dog to a strong sound stimulus) we used three groups of shepherds (242 animals), reared by private individuals and then taken to the kennels (Table 17).

Table 17. Coefficients of Correlation between
the Active Defensive Reaction and the Strength
of the Nervous System

Group No.	Number of individuals, n	Coefficients of correlation $r \pm m_r$	Reliability of correlation r/m_r
1	53	+0.25 ± 0.05	5.0
2	109	+0.12 ± 0.09	1.3
3	80	+0.19 ± 0.10	1.9

The results presented demonstrate the very high degree of correlation between the active defensive reaction and the strength of the nervous system. The dogs with a stronger nervous system are somewhat more likely to possess a less well-marked active defensive reaction than dogs with a less strong nervous system.

In the previous chapter we showed that the manifestation and expression of an active defensive reaction require individual experience in an animal. Dogs reared in conditions of isolation display a less well-marked active defensive reaction than dogs reared in conditions of freedom.

A positive correlation exists between a strong nervous system and the formation of individually acquired skills. It may be postulated that in dogs possessing a stronger nervous system, an active defensive reaction develops more easily under the influence of individual experience than it does in dogs with a weaker nervous system. A strong nervous system inhibits the manifestation of a passive defensive reaction and, conversely, strengthens the expression of an active defensive reaction.

These results are in complete agreement with the researches, cited above, showing that rearing in conditions of freedom, which make possible the training of an active defensive reaction, leads to its strengthening. These same conditions, however, also bring about the overcoming of cowardice. Dogs with a strong nervous system, which more easily form individually acquired skills, can also more easily develop aggressiveness and more easily inhibit cowardice than dogs with a weak nervous system. This is evidently the explanation of the difference in the relationship observed between the strength of the nervous system and the two defensive reactions of behavior.

The findings described in this chapter illustrate, we believe, an important aspect of the teaching of Pavlov on the role of the degree of excitability and of the strength of the nervous system in the formation of behavior. Individual differences in these more important properties and modalities of the nervous system determine the paths along which the formation of the different behavioral acts of the animal proceeds.

CHAPTER 3

THE FORMATION OF ANIMAL BEHAVIOR
IN RELATION TO
CERTAIN ENDOCRINE FACTORS

The relationship, demonstrated in the preceding chapter, between the formation of behavior and the fundamental functional properties of the nervous system, namely its strength and excitability, indicates that all the factors (external and internal) which alter the degree of excitability and the strength of the nervous system must thereby also influence the animal's behavior.

The endocrine glands have a particularly great influence on the fundamental properties of the nervous system. It may therefore be expected that through the intermediary of a change in the functional state of the nervous system, endocrine factors must consistently influence behavior formation in animals.

Numerous investigations have shown that the influence of the thyroid gland on the excitability of the nervous system is especially marked. Large doses of dried thyroid gland, when given to dogs, cause a "collapse" of nervous activity during the first few days, which is followed by a period of increased excitability (Zavadovskii and Zak, 1928; Zavadovskii, Azimov, and Zakharov, 1929; Zavadovskii, Zakharov, and Zolotov, 1929); small doses of thyroidin lead only to an increase in excitability.

Later work has in general confirmed these views. Crisler, Booher, van Liere, and Hall (1933) described an increase in the excitability and the conditioned reflex secretion of saliva in dogs receiving a thyroid gland preparation. Kleitmann and Titelbaum (1936) showed that slight hyperthyroidization leads to an improvement in differentiation and to an increase in the conditioned reflexes.

Tonkikh (1939) showed that an increase in excitability, tachycardia, and dilatation of the pupils accompanied the hyperplasia of the thyroid gland resulting from the administration of pituitary extract. Anderson (1941) observed a marked increase in the passive defensive reaction and in the excitability of a dog receiving thyroid extract.

Work of great interest was done by Petrova (1945), who studied the influence of hyperthyroidization on dogs with different types of nervous activity. Her experiments showed that excessive thyroid administration to dogs of strong type led to an intensification of the process of stimulation and to the termination of an hypnotic state. In dogs of a weak type the administration of thyroidin depressed conditioned reflex activity and carried the nervous system beyond the limits of its working capacity. As a result of hyperthyroidization, a phobia reappeared in one dog which had previously recovered from the phobia.

Similar findings were obtained by Konge (1956), who established that the administration of small doses (0.006–0.027 g/kg body weight) of thyroidin strengthens and concentrates the processes of excitation and inhibition in the cerebral hemispheres. In some dogs, however, doses of thyroidin of 0.018 g/kg caused a decrease in the magnitude of the conditioned reflexes. Hence this author also concludes that the change in the nervous processes in the cerebral cortex under the influence of thyroidin are dependent on the type of the dog's nervous activity.

In experiments by Baranov, Speranskaya, Tendler, and Mityushov (Baranov et al., 1954; Speranskaya et al., 1955) the administration of small doses of thyroidin (0.002 g/kg body weight) to dogs caused a decrease in the combined magnitude of the effects of positive conditioned reflexes, the disinhibition of differentiation, and the alternation of phases and other disturbances of higher nervous activity, characteristic of weakening of the processes of stimulation and inhibition.

According to Pugachev (1953, 1954), thyroidin in doses of 0.05 g has an exciting action on pigeons of an excitable type of nervous system; in pigeons with a weak type of nervous system the conditioned reflex activity is completely suppressed. Large doses of thyroidin (0.4–0.8 g) caused a biphasic change in the conditioned reflex activity of pigeons: a decrease, followed by an increase in excitability. In pigeons of inhibited type, larger individual doses of thyroidin led to the development of limiting inhibition with alternating phases of reflex activity.

The investigations cited thus show that hyperthyroidism leads to an increased excitability of the nervous system, which is expressed by a strengthening of the motor activity of dogs, by changes in their conditioned reflex activity (this change is dependent on the type of nervous system), and by an increase in the manifestation of behavioral acts such as the passive defensive reaction, or of phobias.

The first researches into the influence of removal of the thyroid glands on the higher nervous activity and behavior of animals were undertaken in Pavlov's laboratory. Val'kov (1925) observed the disturbance of the conditioned reflex activity in a dog from which the thyroid gland had been removed. After operation, all the dog's conditioned reflexes failed to regain their normal stability, and apparently had to be developed afresh. Pavlov describes a dog after thyroidectomy by Val'kov as an animal with very low excitability of the cerebral cortex. Later workers reached the same conclusion (Azimov, 1927; Zavadovskii and Zlotov, 1929; Anderson, 1941), showing that removal of the thyroid gland leads to a significant decrease in the excitability of the cerebral cortex. Anderson observed a weakening of the active defensive reaction in one of his experimental dogs after thyroidectomy.

Work on other animals gave similar results. Liddell (1925) extirpated the thyroid gland in sheep and showed that consistent changes take place in their behavior. Whereas in normal animals loud noises (gunfire, for example) cause alarm, thyroidectomized animals hardly react at all to these stimuli. These behavioral changes take place against a background of a considerable lowering of the motor excitability of the animal, which develops after thyroidectomy (Liddell and Simpson, 1925). Following thyroidectomy, there develop considerable disturbances of the conditioned reflex activity of sheep taught to pass through a very simple maze. Administration of small doses of thyroid gland to thyroidectomized animals restored their normal nervous activity (Liddell, 1925).

Investigations carried out by Kunde and Neville (1930) in rabbits showed that the reflex of movement of the skin in response to tactile stimulation, which disappears after thyroidectomy, reappears after oral administration of a thyroid gland preparation to the animal.

Obvious changes were found in the behavior of chicks, depending on the dose of thyroid gland which they received. Zavadovskii and Rokhlina (1927) showed that hyperthyroidization in chicks causes significant changes in both conditioned and unconditioned reflex activity. Whereas large doses of thyroidin lead to the collapse of nervous activity, small doses cause a slight increase in excitability and lead to a clearer pattern of conditioned reflex activity.

The investigations of Liberfarb (1928) showed that the unconditioned reflex movements of the gizzard in chicks are greatly depressed after thyroidectomy. Administration of a thyroid gland preparation restores the lost movements of the gizzard.

Some interesting observations were made by Larionov. As Larionov and Berdyshev (1933) point out, the period of molting, which follows immediately after the egg-laying period, is accompanied by a considerable increase in excitability. At this time increased timidity is also observed (personal communication from Larionov). The investigations of Larionov and Berdyshev (1933) showed that the thyroid function is increased during the molting period in chicks. It is evident that hyperfunction of the thyroid gland, causing an increase in the excitability of the nervous system, also leads to an increase in timidity.

Investigation of the role of the thyroid gland in the ontogenesis of the behavior of birds was conducted by Vasil'ev (1941). Removal of the thyroid gland from young birds (mainly of the crow family) led to retardation in the development of their behavior, which remained at the fledgling stage.

Our investigations (Krushinskii and Kabak, 1947) showed that during hypofunction of the thyroid gland, caused by administration of methylthiouracil, there was a decrease in the motor activity of rats, which was restored after the drug was discontinued. The administration of thyroxin to rats in which the motor excitability had been lowered by methylthiouracil restored their excitability, in spite of the continued administration of methylthiouracil. Gunin (1952) observed drowsiness in dogs after administration of anti-thyroid preparations. After the administration of methylthiouracil to animals, Tendler (1952) and Komissarenko, Buiko, Gluzman, and Teplitskaya (1957) noted weakening of the processes of excitation and inhibition, with the development of parabiotic phases.

The investigations cited above provide evidence that hyperthyroidization causes an increase in the excitability of the nervous system and a strengthening of the manifestation of reflex reactions. On the other hand, extirpation of the thyroid gland or the blocking of its function leads to a decrease in the excitability of the nervous system.

We carried out investigations in dogs (Krushinskii, 1938a) for the purpose of discovering the action of the thyroid gland hormone on the manifestation of unitary reactions of behavior and the degree of excitability of the nervous system.

Experiments to determine the influence of hyperthyroidization and thyroidectomy on the manifestation and expression of the unitary reactions of behavior were carried out on ten dogs. Observations were made on the animals for 2–3 months before the experiment began. During this time all the experimental dogs but one were very stable in their behavior. Their excitability was measured by means of a pedometer, which was attached to the dogs during twelve night hours (from 7 P.M. until 7 A.M.). The behavior of each dog after hyperthyroidization or thyroidectomy was compared with its normal behavior.

In Table 18 and Fig. 10 we give the results of the experiments to study the influence of thyroidization on the motor activity and defensive behavior of dogs.

Table 18. The Effect of Thyroidin on the Motor Activity and Defensive Behavior of Dogs

Name of dog, time of giving and dose of thyroidin	Before hyperthyroidization		After hyperthyroidization	
	Mean of 5 pedometer readings	Defensive behavior	Mean of 5 pedometer readings	Defensive behavior
Ada From March 28 to May 6 received 96 g thyroidin	4800 (2500–8000)	weakly expressed passive defensive reaction	19,400 (9700–25,000)	strengthening of the passive defensive reaction
Baikal From June 2 to July 25 received 336.5 g thyroidin	3100 (1400–4500)	weakly expressed active defensive reaction against a background of passive defensive behavior	18,900 (6000–25,000)	strengthening of expression of active defensive component (lasting 1.5 months)
Lyumb From March 29 to July 9 received 508.5 g thyroidin	3500 (3000–4300)	very slight active defensive reaction	5400 (3600–75,000)	slight strengthening of active defensive reaction
Liza From August 12 to October 15 received 675 g thyroidin	3260 (1500–6000)	ill-defined active defensive reaction	33,000 (28,000–38,000)	marked strengthening of active defensive reaction
Yarok From August 11 to August 19 received 105 g thyroidin	–	well-defined active defensive reaction	–	considerable strengthening of active defensive reaction
Al't From November 28 to January 2 received 787.5 g thyroidin	3040 (1400–4400)	active defensive reaction (only in the form of baring teeth) (Fig. 10A)	5930 (4400–7500)	strengthening of active defensive reaction (began to growl) (Fig. 10B)

It will be seen in Table 18 that hyperthyroidization of the dogs, leading to an increase in their excitability, also strengthened the expression of their defensive behavioral reactions.

A

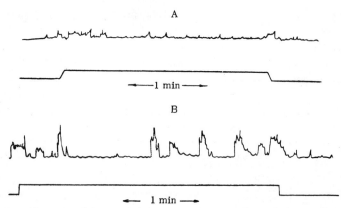

←—1 min—→

B

Fig. 10. Active defensive behavior before (A) and after (B) administration of thyroidin. (Recording of barking.)

In Table 19 we present data showing the influence of thyroid-ectomy on the motor activity and defensive behavior of dogs. The results in Table 19 show that thyroidectomy leads to a decrease in excitability (in 3 of 5 individuals) and a weakening of defensive behavior. Although the administration of thyroidin led to a more noticeable change in the behavior of dogs than removal of the thyroid gland, nevertheless this investigation demonstrates the influence of the functional state of the thyroid gland on the excitability and the defensive behavior of dogs.

Reports in the literature and our own investigations have shown that the thyroid gland influences animal behavior, and that this influence is exerted through a change in the excitability of the nervous system. A hyperthyroid state leads to an increase in excitability, and thereby to a strengthening of the expression of the unitary reactions of behavior. Conversely a hypothyroid state, which is accompanied by a decrease in the excitability of the nervous system, leads to a weakening of the expression of unitary reactions of behavior.

The relationship which we have studied between the thyroid gland and the defensive behavioral reactions reveals only one of the many factors influencing the excitability, and hence the behavior of the animal. Changes in the activity of the thyroid gland, in turn, depend on many conditions and processes taking place in the body, and these, through changes in the functional state of the gland, must have an influence on the formation of the animal's behavior.

Other glands of internal secretion influencing the functional state of the nervous system and the behavior of animals is the gonad.

Table 19. Effect of Thyroidectomy on the Motor Activity and Defensive Behavior of Dogs

Name of dog	Before thyroidectomy		After thyroidectomy*	
	Mean of 5 pedometer readings	Defensive behavior	Mean of 5 pedometer readings	Defensive behavior
Ada	19,400 (9700–25,000)	passive defensive reaction	7380 (4000–10,000)	weakening of passive defensive reaction
Ledi	1,980 (1200–2700)	active defensive reaction	2300 (1250–2700) 2.5 mos. after operation	almost complete disappearance of active defensive reaction
			1700 (1100–2700) 11 mos. after operation	active defensive reaction restored
Vernyi	4,540 (2500–9000)	active defensive reaction	4100 (3000–5500)	active defensive reaction unchanged
Sil'va	3,700 (2700–5000)	active and passive defensive reactions	1600 (1000–2200)	no significant change in defensive behavior
Asta	6,500 (5000–8600)	active and passive defensive reactions	1800 (1250–3000)	considerable weakening of active and passive defensive reactions

*In one dog (Ada) the thyroid gland was removed two days after cessation of the hyperthyroidization.

Most of the work on the study of sex differences in the degree of excitability (as reflected by the motor activity, measured by means of activity wheels) has been undertaken in rats. Slonaker (1912), in one of his early papers, showed that female rats are more active than male rats. These findings were subsequently confirmed by Hitchcock (1925) and others. Wang (1923) and Slonaker (1925) later found that female rats display increased spontaneous activity during estrus. Spaying of female rats leads to a considerable decrease in motor activity (Wang, 1925; Slonaker, 1930); the same occurs in males (Hoskins, 1925, 1925a; Gans, 1927; Lee and Buskirk, 1928; Slonaker, 1930; Heller, 1932).

In contrast to these findings there is the work of Tuttle and Dyckshorn (1928), who conclude that castration of male and female rats is not reflected in their "spontaneous" activity. Although in the majority of investigations the results obtained indicate a decrease in the motor activity of castrated males when measured by means of activity wheels, it was found that if the motor activity is measured by other methods the results of these measurements differ from those obtained by the use of activity wheels (Campbell and Sheffield, 1953; Eayrs, 1954).

The age at which castration is carried out is evidently important. Richter (1933) found that males castrated at the age of 3–4 days, although they were less active than control animals, were nevertheless more active than rats castrated later in life. Wang, Richter, and Guttmacher (1925) showed that the transplantation of testicles into castrated males leads to an increase in their motor activity. The same investigation clearly showed that the motor activity of females is increased during the period of estrus. The injection of folliculin into spayed females, producing estrus, leads to an increase in motor activity (Bugbee and Simond, 1926). Richter and Wislocki (1928) showed that the motor activity of rats of both sexes diminished after castration and is restored after the gonads are replaced by transplantation. Transplantation of the testicles has less effect on the motor activity than transplantation of the ovaries. Similar results were obtained by Slonaker (1927) in old menopausal rats with a decreased motor activity; the injection of folliculin led to an increase in their activity.

Results slightly different from those cited above were obtained by Heller (1932). He was unable to find any significant change in the motor activity of male rats when they were given injections of male sex hormone.

The researches which we have mentioned show that the motor activity of rats is related to the sex hormones; it is especially influenced by the female sex hormone. The normal motor activity of males evidently requires a definite level of male sex hormone in the body.

Investigations in pigs showed that the spontaneous activity of these animals increases in females during estrus (Altmann, 1939). The activity is maximum at the end of estrus. Before estrus a period of perceptible depression of excitability of the animal is observed.

Humphrey and Warner (1934) compared the excitability of 174 male and 172 female trailer dogs. They found that the average excitability of the females was rather higher than that of the males. In our investigations we measured the excitability of dogs of both sexes in six groups of German shepherds, numbering 390 altogether (Table 20).

Table 20. Difference in the Degree of Excitability of Male and Female Dogs*

Group No.	♂♂		♀♀		Difference	Difference in favor of
	n	Excita-bility	n	Excita-bility	M(diff)	
1	21	5.24 ± 0.25	35	5.43 ± 0.19	0.19 ± 0.32	♀♀
2	63	5.29 ± 0.16	36	5.44 ± 0.19	0.15 ± 0.26	♀♂
3	23	5.57 ± 0.15	31	5.39 ± 0.15	0.18 ± 0.20	♂♂
4	35	5.80 ± 0.14	26	5.42 ± 0.24	0.38 ± 0.28	♂♂
5	37	5.35 ± 0.18	32	5.06 ± 0.19	0.29 ± 0.26	♀♀
6	26	5.92 ± 0.14	25	5.80 ± 0.21	0.12 ± 0.25	♂♂

*For biometric analysis, the total number of dogs was divided (according to the pedometer readings) into seven classes. The analysis was made in accordance with the classes, and not the pedometer readings.

It will be seen from Table 20 that no significant difference could be found between the excitability of the males and females. Subsequently the degree of excitability was determined (jointly with Fless) in two further groups of German shepherds. In both groups (128 dogs altogether) the males were slightly more excitable than the females, but the difference was not significant.

It is thus difficult to conclude from the findings described above that any essential sex-determined difference exists in the excitability of dogs.

The influence of the sex hormones on conditioned reflex activity has been studied by several researchers. In investigations carried out in rats it was impossible to establish any obvious influence of the sex hormones on the course of formation of individually acquired skills.

Tsai (1930) came to the conclusion that completely castrated male rats make more mistakes and take longer to pass through a maze than males from which only one testicle has been removed. The latter, in turn, do not learn as well as normal males.

Tsai's findings were not confirmed by Commins (1932), who found no difference in the ability of normal and castrated male rats to learn. Boll (1926) also could not detect any essential difference in the learning capacities of rats during rut and during a period of sexual quiescence.

Beach (1956) found obvious differences between normal and castrated males in the formation of a motor habit of running from one compartment to another in which there were females in heat. On the average the normal males run four times as fast as those castrated. Injection of male sex hormone fully restored the speed of the castrated males. The differences obtained between the normal and castrated males in the formation of a habit could undoubtedly be explained by the specificity of the reinforcement used (females in heat).

Much clearer results have been obtained in dogs. Arkhangel'skii (1931, 1937) showed that removal of the male gonad has a considerable effect on the conditioned reflex activity of dogs. Castration leads to extreme weakening of the functional activity of the cortex, as shown by its rapid fatigability, by signs of weakening of the strength of the nervous system and of a diminution of its working capacity, and by increased difficulty in the formation of conditioned reflexes. After castration, both excitation and inhibition are weakened. The effect of castration is especially strong in young animals, in which it causes considerable disturbances of conditioned reflex activity. A long time after castration some restoration of the conditioned reflex activity of the experimental dogs was observed, but it never regained its preoperative level.

Petrova (1936, 1937) studied the influence of removal of the male gonad of dogs on their conditioned reflex activity. Her conclusions are in agreement with the results of Arkhangel'skii: castration led to weakening of the activity of the cerebral cortex. The importance of the type of the animal's nervous system is made clear by Petrova's work. In individuals with a strong nervous system the strength of the nervous processes is more or less restored after operation. After castration, animals with a weak nervous system may at first even begin to "work" a little better under stereotyped conditions, but subsequently a profound depression develops and they completely refuse to work. Castration of a puppy led to considerable weakening, primarily of inhibition: when this came up against the process of excitation a severe and prolonged neurosis developed.

Work on castration thus showed that absence of the male sex hormone leads to the weakening of the activity of the cerebral cortex.

Implantation of testicles (Arkhangel'skii, 1927, 1937) into adult males led to characteristic changes in the conditioned reflex behavior. After implantation the normal relationships between the processes of excitation and inhibition are disturbed; the former begin to predominate over the latter. Arkhangel'skii explains this phenomenon not by an increase in the process of stimulation, but by a weakening of inhibition. After transplantation of testicles into castrated males, the working capacity of their cerebral cortex approximated to that of unoperated animals, but it was not completely restored.

In a series of researches the role of the hormones of the female gonads was studied. Arkhangel'skii (1937) cites the results of Gekker's experiments on castration of female dogs, and concludes that conditioned reflexes are not affected by spaying. Similar results were also obtained when ovaries were transplanted into females (Gekker, 1931). From all these findings, Arkhangel'skii considers that in females the nervous activity is influenced to only a slight degree by the ovarian hormones.

However, many workers (Artem'ev, 1939; Usievich, Artem'ev, Alekseev, and Stepanov, 1938) have concluded that spaying of females leads to approximately the same consequences as are observed after castration of the males—a weakening of the processes of stimulation and inhibition—and, moreover, that the process which is relatively weaker in the particular animal is usually more severely affected. The findings of these workers thus are opposed to the views of Arkhangel'skii and Gekker concerning the absence of a relationship between the higher nervous activity and the function of the ovaries in dogs.

Examination of the facts from which Usievich and Artem'ev drew their conclusions that the ovaries do influence the higher nervous activity, indicates that although in females, after spaying, the conditioned reflex behavior does show abnormalities, these changes are not very great, and are considerably less prominent than the changes in behavior observed after castration in males. The very insignificant relationship between the higher nervous activity of dogs and the female gonad in comparison with that of the male gonad evidently led Arkhangel'skii and Gekker to the opinion that the internal secretion of the ovary exerts almost no influence on the cerebral cortex.

We must also point out that the results of experiments in which the sex glands are implanted, without subsequent control histological examination of the implants (which was not done in the work cited above) are not quite convincing.

Anderson (1941) also concludes that both male and female sex glands influence the conditioned reflex activity of dogs. He castrated three dogs (two males and one female). The effect in all cases was the same: the magnitude of the conditioned reflexes was slightly diminished.

The investigations of Kreps (1924) demonstrate the influence of an increased production of female sex hormone on the conditioned reflex activity. During estrus he observed characteristic waves of increased excitation and weakened inhibition, alternating with a state of depression.

The researches cited above thus showed that the importance of sex hormones evidently differs from one animal to another. Whereas in rats their influence on conditioned reflex activity is only slight, in dogs their influence is beyond doubt. In these animals the presence of sex hormones is essential for the maintenance of the strength of the nervous system at its normal level, and moreover the male sex hormone evidently has a greater influence on the strength of the nervous system than does the female.

Sexual dimorphism in the formation of individually acquired skills was first investigated in rats. Ulrich (1915) found that when rats were taught in a "problem box" or in a maze, the males learned the skill more quickly and went through the maze faster than the females. The sex difference was not so clearly apparent in the young (25-day-old) and old (300-day-old) rats.

Akimov (1928, 1930) studied the sex difference in the ability of rats to learn, and concluded that males acquire skills connected with exteroceptive sensation more easily than females; females acquire skills connected with proprioceptive sensation more easily than males. The more recent investigations of Tryon (1931) and of McNemar and Stone (1932) showed that when rats were taught in various types of maze the males learned more rapidly than the females, and the females made slightly more mistakes than the males. The conclusion from all these investigations is that males learn more easily than females.

These results do not tally with the work of Corey (1930), who concluded from his experiments that female rats learn more quickly and make fewer mistakes than males. It is difficult to account for this divergence from the findings of other workers.

Most of the evidence thus goes to show that males show a better ability to learn than females. In some cases, however, the analyzer playing the leading part in the formation of the skill may be important. It may be considered that the analyzers connected with proprioceptive function are more highly developed in females than in males.

We carried out investigations to study the differences between male and female dogs in the formation of individually acquired skills and in the strength of the nervous system. Let us begin by examining the sex differences in the ability of the dogs to undergo training (in antitank and communications work during wartime). The results of such training of males and females are shown in Table 21.

Table 21. Formation of an Individually Acquired Skill in Male and Female Dogs

Group No.	Mean values of indices of success in training (on a system of points)		Difference	Difference in favor of
	♂♂	♀♀		
1 (communications)	3.43 ± 0.24	3.69 ± 0.19	0.26 ± 0.32	♀♀
2 (AT1)	3.52 ± 0.14	3.22 ± 0.17	0.30 ± 0.22	♂♂
3 (AT1)	3.39 ± 0.26	3.94 ± 0.18	0.55 ± 0.32	♀♀
4 (AT2)	4.09 ± 0.13	3.62 ± 0.17	0.47 ± 0.22	♂♂

It will be seen from the figures in Table 21 that, at first glance, there is no consistent difference between the trainability of males and females. In groups 2 and 4 the males were trained slightly better than the females, but in 1 and 3, on the other hand, the females gave better indices, and moreover all the differences that were observed lie within the limits of probable error. From a more detailed analysis, however, it is possible to establish sex differences in the formation of individually acquired skills.

It was pointed out in Chapter 2 that the conditions of training of the four groups of dogs investigated were of different degrees of difficulty. On the basis of the physical strength of the stimuli applied, and the strain (in relation to the periods of instruction), these conditions must be classed as difficult for group 2, moderately difficult for groups 1 and 4, and easy for group 3. Since the dogs of groups 2 and 3 were trained for the same job (AT1, antitank work), it is possible to compare the training ability of dogs of both sexes in easy and difficult conditions. The results of such a comparison are shown in Table 22.

Table 22 shows that there is no difference in the success of formation of an individually acquired skill between the males in the two groups for all practical purposes, whereas the difference between the females is considerable, and is statistically significant. Females learn a skill less readily when trained under difficult

Table 22. Formation of a Skill in Males and Females in Easy and Difficult Conditions

Sex	Mean assessment (points)		Difference Md ± md	Significance of difference Md/md
	Group 3, easy conditions	Group 2, difficult conditions		
Males	3.39 ± 0.26	3.52 ± 0.14	0.13 ± 0.30	0.43
Females	3.94 ± 0.18	3.22 ± 0.17	0.72 ± 0.25	2.88

conditions than when conditions are easy. In difficult conditions of the formation of an individually acquired skill, when great concentration is required on the part of the animal's nervous system to enable it to withstand strongly acting external factors, greater stability of the formed skill is observed. In females, difficult conditions interfere with the course of formation of the individually acquired skill, which indicates that the nervous system of females is weaker than that of males.

Our analysis (Krushinskii, 1946) of material relating to the strength of the nervous system in male and female dogs of the kennels of the Pavlov Institute of Physiology in Koltushy, confirmed the conclusion that the nervous system is stronger in males than in females (Table 23).

Table 23. The Strength of the Nervous System in Dogs of Different Sex *

Strength of the nervous system	♀♀	♂♂	Σ
Very strong	0	5	5
Strong	7	16	23
Weak variant of the strong type	3	4	7
Strong variant of the weak type	5	1	6
Weak	7	5	12

*The type of nervous system of the dogs to be compared was determined by members of the staff of the Institute of Physiology in accordance with the accepted standard (Timofeeva, 1941).

Calculation of the criterion of correlation (x^2), which equals 10.29, shows that the difference between males and females according to the strength of their nervous system (on the basis of the probability of correlation p/x^2 being less than 0.05) is significant. Males possess on the average a stronger nervous system than females.

In addition to the above results obtained on the basis of data concerning the formation of an individually acquired skill, we compared the results of determinations of the reaction of male and female German shepherds (356 animals) to the action of a strong sound stimulus (a rattle). The results were as follows: for males $M = 6.25 \pm 0.19$; for females $M = 6.14 \pm 0.19$ (an 8-point scale was used to mark the dog's ability to withstand the sound of the rattle). The difference (Md) was 0.11 ± 0.28 in favor of the males. The significance of the difference (Md/md) was 0.39. Thus, no statistically significant difference could be established between the males and females of the group of shepherds investigated in relation to their "resistance" to the sound of the rattle.

We believe that we were unable to detect a difference in the strength of the nervous system because the strength of the stimulus used (the rattle) was inadequate. As we pointed out above, when the dogs were trained in easy conditions, it was also impossible to detect this difference. In difficult conditions, however, in which the nervous system of the dogs was subjected to extreme strain, the difference was quite clear. This was confirmed by the results of an investigation carried out jointly with Fless on trailer dogs (63 animals), in which a louder rattle was used, together with a siren. The action of these stimuli was withstood by 81–83% of the male dogs, but by only 35-40% of the females. At the same time, trials with large doses of caffeine in the same animals revealed no sex difference in the strength of the nervous system, evidently as a result of the inadequacy of the dose used (0.8 g of caffeine sodium benzoate).

We may conclude from these results that the nervous system of male animals is stronger than that of females. However, this difference is only manifested when strong stimuli act on the nervous system.

Let us now pass to the examination of sex differences in the manifestation and expression of unitary reactions of behavior. Without dwelling on the description of forms of behavior which are specifically connected with the gonads, and which may undoubtedly be regarded as secondary sexual characteristics (for example, the song or cry of various species of male birds and other forms of mating behavior, reactions directed toward copulation, and so on), let us examine the relationship between the formation of unitary reactions and sex, using defensive behavior as an example.

Although these unitary reactions in a number of cases depend on the functional state of the gonads, they must not always be classed as secondary sexual characteristics. The stimulus in relation to which the defensive reaction of the animal is manifested is also of importance. The active defensive reaction of males against each other undoubtedly depends on the functional state of the gonads.

Castration of cocks greatly weakens their aggressiveness (Zavarovskii, 1922). Ulrich (1938) showed that castration weakens the aggressiveness of certain male rats. The time of castration is very important. When carried out before the onset of sexual maturity, castration almost completely abolishes aggressiveness; castration of sexually mature animals does not abolish it. The aggressiveness of male rats and mice, once weakened by castration, is restored after the injection of male sex hormone (Riege, cited by Seward, 1945; Beeman, 1946). J. G. Seward (cited by Seward, 1945) found a clear sex difference in the expression of the active defensive reaction in male and female rats. In the former the reaction was much more pronounced. Injection of a preparation of male sex hormone (androgen) led to an increase in aggressiveness; injection of female sex hormone (estrogen) had no appreciable effect.

An interesting example of the role of the functional state of the gonads in the formation of the active defensive behavior of dogs is mentioned by Tinbergen (1955). His observations showed that Eskimo dogs exhibit a well-marked territorial sense within the limits of the Eskimo settlements. The dogs of the settlement are separated into a few small colonies or packs, each of which lives in a definite part of the settlement. The appearance of a dog from another pack on "foreign" territory at once causes all the members of the colony to display an active defensive reaction against it, and they chase the "foreign" dog out of their own territory. Young dogs do not belong to any one colony, but wander all over the settlement, in spite of persecution. After the first mating, however, the behavior of the dog changes. It at once joins a certain pack, it ceases to appear on "foreign" territory, and it chases dogs from a "foreign" colony off its "own" territory. It can be seen from this example how the functional state of the gonads shapes the active defensive reaction of the dog's behavior, which in turn plays an important role in the formation of the pack.

The passive defensive reaction is much less obviously dependent upon the functional state of the gonads. As Anderson (1940a) showed, the cowardice of female dogs is less marked during the period of rut. Injection of female sex hormone weakened the cowardice of normal and spayed females (E. Anderson and S. Anderson, 1940b). However, castration of neither males nor females leads to a change in their passive defensive reaction (Anderson, 1940c). From these experiments, Anderson concluded that an increased amount of female sex hormone circulating in the body weakens manifestations of cowardice; absence of sex hormone, however, does not increase cowardice.

It thus follows from the literature that the male sex hormone has an important influence on the manifestation and expression of the active defensive reaction in the males of various animals. It was not possible to establish any clear relationship between the passive defensive reaction and the female sex hormones, although there are reports that this reaction is weakened by them.

Table 24. Passive Defensive Reaction in Males and Females

Group of dogs	n	Mean indices of the passive defensive reaction (in points)		Difference Md ±md	Difference in favor of	Significance of difference Md/md
		♂♂	♀♀			
Mongrels (kennels of the Pavlov Institute of Physiology)....	72	2.17 ± 0.29	2.39 ± 0.32	0.22 ± 0.42	♀♀	0.52
German shepherds belonging to individuals	62	0.62 ± 0.23	1.55 ± 0.16	0.93 ± 0.26	"	3.5
German shepherds from kennels	58	3.13 ± 0.38	3.48 ± 0.31	0.35 ± 0.44	"	0.8
Airedale terriers belonging to individuals	41	0.18 ± 0.15	0.33 ± 0.16	0.15 ± 0.22	"	0.7
Airedale terriers from kennels	111	0.50 ± 0.12	1.0 ± 0.18	0.50 ± 0.21	"	2.3
German shepherds from a training school	238	1.38 ± 0.06	1.58 ± 0.08	0.20 ± 0.10	"	2.0

Although our findings concerning the manifestation and expression of defensive reactions in dogs (in relation to man) reveal the existence of slight differences in their expression in different sexes, they nevertheless show that these reactions cannot be regarded as definite secondary sexual characteristics.

Our investigations into sex differences in the passive defensive reaction of dogs were carried out in 582 dogs (284 males and 298 females). In connection with differences in the breeds of the dogs and the conditions of their upbringing, the total number of animals was divided into six groups (Table 24).

It will be seen from Table 14 that in all six groups the passive defensive reaction is slightly more marked in the females than in the males (although in three groups the difference is not statistically significant). More recent investigations carried out in German shepherds of the trailer-dog service gave similar results. In a group of these dogs that was studied (61 animals), 31.6% of the females and only 7.1% of the males possessed a passive defensive reaction.

The investigation of the sex difference in the manifestation of the active defensive reaction was carried out in 421 dogs (Table 25).

Table 25. Active Defensive Reaction in Males and Females

Group of dogs	n	Mean indices of the passive defensive reaction (in points)		Difference Md±md	Difference in favor of	Significance of difference Md/md
		♂♂	♀♀			
German shepherds belonging to individuals	62	2.61 ± 0.10	2.47 ± 0.11	0.14 ± 0.14	♂♂	1.0
German shepherds from kennels	47	2.05 ± 0.18	1.96 ± 0.18	0.09 ± 0.18	"	0.5
Airedale terriers belonging to individuals	37	2.53 ± 0.14	2.36 ± 0.17	0.17 ± 0.22	"	0.8
Airedale terriers from kennels	96	1.79 ± 0.13	1.67 ± 0.12	0.12 ± 0.17	"	0.7
German shepherds from a training school	42	2.06 ± 0.22	2.00 ± 0.21	0.06 ± 0.30	"	0.2
German shepherds from a training school	86	2.39 ± 0.13	2.27 ± 0.19	0.12 ± 0.22	"	0.5
German shepherds from a training school	51	2.41 ± 0.15	1.74 ± 0.23	0.67 ± 0.26	"	2.6

It is clear from Table 25 that in all the groups of dogs investigated, the males possess a slightly more marked active defensive reaction than the females. Although this difference is very small, and is in all cases statistically insignificant (only in one group does it approximate to the limits of statistical significance), nevertheless we do not regard it as an accidental occurrence.

Among trailer dogs which we studied later (61 animals) the proportion of dogs with an active defensive reaction and the degree of its expression were also higher in males than in females.

The findings described in this section thus indicate that differences exist between the fundamental properties and modalities of the nervous activity and behavior of males and females. Sex dimorphism in nervous activity manifests itself differently in different animals. In dogs, consistent differences are observed between the behavior of males and females: males possess a stronger nervous system, which enables them to form individually acquired skills more successfully under frustrating and difficult situations; in males the passive defensive reaction is less prominent, and the active possibly more so. No differences could be found in the degree of excitability between dogs of the two sexes.

We consider that the difference between the behavior of dogs of the two sexes is based upon a difference in the strength of the nervous system. The remaining differences in behavior are a consequence of this. Sex differences in the expression of unitary reactions (passive and active defensive) are insignificant, because they reflect only indirectly the difference in the degree of the strength of the nervous system of males and females.

The findings described in this chapter illustrate some of the ways in which endocrine factors influence the formation of animal behavior. The thyroid gland affects the threshold of excitability of the nervous system. One of the paths of action of the gonads is via a change in the strength of the fundamental nervous processes, and hence it affects the formation of the behavior of animals.

THE IMPORTANCE OF THE FUNCTIONAL STATE OF THE NERVOUS SYSTEM IN THE MANIFESTATION OF THE PATHOLOGICAL REACTIONS OF THE BODY

In the book "The Reflexes of the Brain" (1863) Sechenov gave the first clear indication that the active excited state of the brain is maintained by the sum total of stimuli received by the organs of the senses.*

This concept was subsequently proved in Pavlov's laboratories. It was shown that the maintenance of the normal tone of the brain requires a constant influx of impulses from the receptor apparatus.

The new notion of the conduction of excitation in the analyzer highlights the tremendous importance of afferent impulses in maintaining the state of excitation of the cerebral cortex—impulses which are transmitted along not only specific but also nonspecific paths. There is a considerable volume of evidence that the flow of afferent impulses arises from specific conduction pathways in the reticular formation of the brainstem and thalamus, from whence they exert an activating influence on the cerebral cortex along ascending, nonspecific pathways (see Lorente de Nó, 1951; Brazier, 1955; Sokolov, 1958). In various functional diseases of the brain, when the principal symptom is a state of cerebral excitation, a method of treatment is to provide complete rest for the nervous system by means of a decrease in the number and strength of the stimuli in the external environment. It is therefore of very great importance to make a detailed study of the action of strong external stimuli on the nervous system, of the functional changes which take take place within it, and of the system of "defense" which it employs under these circumstances.

In the course of an animal's life it is exposed to the action of strong, and sometimes excessively strong stimuli, which cause profound changes in its functional state. In order to study the laws governing the development of any human disease, and to develop methods of controlling it, it is of great value to have an experi-

---

*I. M. Sechenov, Selected Works (Izd. VIEM, Moscow, 1935), p. 235.

mental model of that disease. Pavlov considered that "the experimental study in animals of pathological changes in the fundamental processes of nervous activity will lead to a physiological understanding of the mechanism of the bulk of neurotic and psychotic symptoms, whether existing separately or combined as definite pathological forms."*

The classical researches of Pavlov and his school showed how exceptionally fruitful was the study of experimental neuroses in dogs for the understanding of the physiological mechanisms of both pathological and normal nervous activity. Pavlov pointed out that the "normal level of nervous activity is a state of equilibrium between all the processes which have been described as taking part in this activity. The disturbance of this equilibrium is a pathological state, or disease, so that it often happens that in what is called normal, or what should more accurately be called relatively normal, some disturbance of this equilibrium actually exists."† The creation of experimental models of various diseases of the human nervous system in laboratory animals must extend the investigations of Pavlov and his school in this direction.

A very suitable model for the experimental study of several pathological disturbances of nervous activity is the marked excitation which develops in rats and mice as a result of the action of a strong sound stimulus. This excitation often terminates in a series of gross pathological states: convulsions, catatonia, death from hemorrhagic shock, the development of a chronic motor neurosis, etc.

Marked excitation and convulsions in mice in response to the action of the sound of a bell were obtained for the first time in Pavlov's laboratory by Studentsov (see Vasil'ev, 1924) and Ganike (personal communication from V. K. Fedorov). Donaldson (1924) also mentions the development of marked excitation and convulsions in rats from the action of the sound of the jingling of keys.

The value of this model is that, without administering any drugs to the animal, without using strong stimuli, such as an electric current, and without destroying the integrity of the nervous system, convulsions and epileptiform fits may be obtained in response to the sound of an ordinary electric bell.

After 1938, experimental epileptiform fits ("reflex epilepsy" — Krushinskii, 1949) in response to a sound stimulus began to be

*I. P. Pavlov, op. cit., p. 734.
†Ibid, p. 723.

studied intensively, especially by workers abroad. These investigations were devoted to the study of the role of sex, age, the glands of internal secretion, diet, drugs, ionizing radiation, and other factors on the development of convulsions (see Krushinskii, 1949, 1954).

These investigations, however, did not reveal the physiological mechanisms lying at the basis of this pathological state. In the conclusion to his survey of investigations of convulsions in rats, Finger (1947) writes: "The very interesting field of investigation of animal behavior associated with the study of convulsive fits is so complicated and complex that at the present time no physiological explanation of this phenomenon can be given."

From the very outset of our own investigations we attempted to study those physiological mechanisms which lie at the basis of the excitation reaction of rats in response to the action of a strong sound stimulus (Krushinskii, 1949, 1954; Krushinskii, Fless, and Molodkina, 1950) from the standpoint of the laws of the physiology of nervous activity, established by Sechenov, Vvedenskii, and Pavlov.

Our investigations, which at first were apparently concerned with a special problem of pathology, far exceeded the bounds of the study of reflex epilepsy in rodents. They are concerned with the central problem of higher nervous activity—the interaction between the process of excitation and inhibition and those physiological "defensive measures" used by the body against strong excitation of the brain, which is the cause of a number of severe pathological states, terminating in some instances by death.

Czech physiologists (Bures, 1953, 1953a, 1953b; Servit, 1952, 1955) have also undertaken the physiological analysis of this reaction.

We cause excitation in rats by means of the action of the sound of a loud bell (about 112 decibels). The animal is placed in a chamber measuring $42 \times 26 \times 50$ cm (Fig. 11). Inside the chamber there is an electric bell. The intensity of the sound can be varied between 112 and 70 decibels by changing the voltage of the current by means of a transformer (between 130v and 20v). Conventionally we express the intensity of the sound not in decibels, but by the voltage of the electric current supplying the bell.

Investigations carried out in our laboratory by Semiokhina and Kagan showed that the line of rats used for the experiments was most sensitive to sound stimuli lying in the relatively high frequency range: over 14-16 kc, i.e., at the limit of audibility of the human ear. Lower frequencies (3-4 kc) produce an excitation reaction in only a few rats.

Fig. 11. Apparatus for exposure of rats to sound.

The results obtained in our laboratory are in agreement with those of other workers. In the experiments of Gould and Morgan (1941), for instance, the effectiveness of the sound increased with an increase in its frequency to 40 kcps. Frings and Frings (1952) also found that the highest incidence of fits took place in response to the action of ultrasound.

The motor reaction of the rats is recorded on a kymograph, by means of pneumatic transmission of the oscillations of the movable base of the chamber. At the same time the character of the response reaction is recorded in note form in accordance with a system of points which we have devised (Krushinskii, 1949; Krushinskii, Fless, and Molodkina, 1950), as follows:

0 – absence of motor excitation during action of the sound stimulus for 1.5 – 2 minutes;

1 – motor excitation in response to the action of a sound stimulus (haphazard jumping and running), not culminating in a convulsive fit (Fig. 12);

2 – motor excitation culminating with the animal suddenly passing into a state of stupor and falling on its belly (usually followed by clonic convulsions) (Fig. 13);

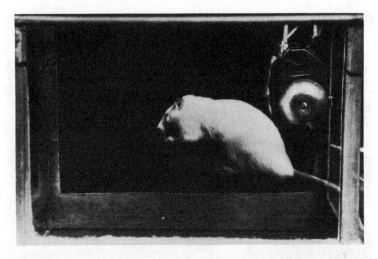

Fig. 12. Motor excitation of a rat during the action of a sound stimulus ("1").

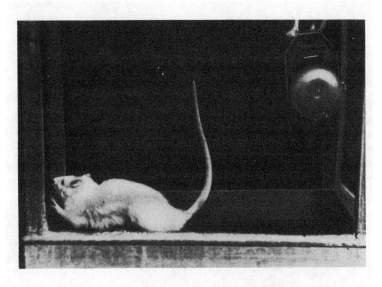

Fig. 13. Convulsive fit with the rat falling on its belly ("2").

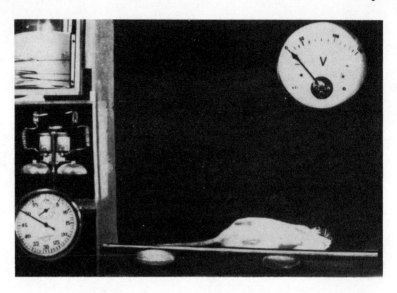

Fig. 14. Convulsive fit with the rat falling on its side. Clonic phase ("3").

Fig. 15. Convulsive fit. Tonic phase ("4").

3-motor excitation culminating with the animal falling on its side, with clonic convulsions (usually of the forelimbs) (Fig. 14);

4-motor excitation culminating with the animal falling on its side, with tonic contraction of all its muscles, and holding its breath for a few seconds (Fig. 15).*

Our investigations showed that the course of the excitation in response to the action of a sound stimulus principally takes the form of one or two waves of motor activity.

Single-wave excitation. Usually 3-5 seconds after application of the sound stimulus intensive motor excitation begins. In most cases at the moment of maximal excitation a convulsive fit develops. Thus from the beginning of excitation to the convulsive fit or to the moment of cessation of the sound stimulus the rat is in a continuous state of excitation (Fig. 16).

Double-wave excitation. From 5 to 15 seconds after application of the sound stimulus excitation begins to develop in the rat, and after 5-10 seconds this suddenly is interrupted and a period of inhibition develops,

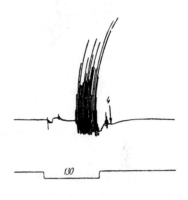

Fig. 16. Motor excitation, in the form of a single wave. The figure (4) marks the position of the convulsive fit (tonic phase). Top curve—recording of the animal's movements. Middle tracing—marker of time of application and cessation of stimulus. Bottom tracing—time marker (interval 1 second).

in the course of which the animal stays motionless in its place. After 10-20 seconds the second wave of motor excitation begins, which terminates in a convulsive fit or continues until the sound stimulus is discontinued (Fig. 17).

After the fits, the rats are often in a state of "aggression." With a characteristic squeak, the animal rushes about the chamber. If a rat in such a state is held in the hand, it will bite. The state of aggression never appears if the rat displayed motor excitation alone, without a convulsive fit. After a very intensive tonic fit, aggression is rarely observed. Both after excitation alone and after a convulsive fit of any degree, a catatonic state is observed in the animals, during which they may adopt curious postures.

*Quantitative variations on either side of these principal types of pathological reactions we assess by plus (+) and minus (−) signs.

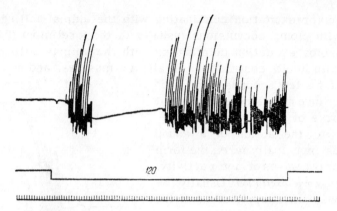

120

Fig. 17. Motor excitation in the form of two waves.

If application of the sound stimulus for two minutes does not cause motor excitation, the rat nevertheless, often shows some evidence of excitation: exophthalmos, vibration of the vibrissae, turns or half-turns in a circle, washing or more active licking of its fur. The last symptom is evidently a characteristic sublimation of pathological excitation. Its appearance is a fairly reliable criterion of the fact that motor excitation will not develop in response to the stimulus used.

From the very first steps of our work we encountered great difficulties: for the experiments we had to select from random populations of rats that were sensitive to the sound stimulus. In the great majority of cases the animals selected exhibited a very inconstant reaction in response to the sound stimulus. Usually a rat, having once or twice shown an excitation reaction, when tested subsequently no longer reacted to the sound stimulus. In conjunction with Molodkina, we therefore in 1947 undertook selection work to isolate a line of rats sensitive to the sound stimulus. For this purpose we selected from various nurseries rats which were most sensitive to the sound stimulus, and used them for crossing.

From its beginning with the most sensitive rats, after a few generations selection greatly increased the incidence of manifestation and the degree of expression of the pathological reaction of excitation (Fig. 18). We have now isolated a line of rats highly sensitive to the action of the sound stimulus, which in 98-99% of cases (in comparison with 10-15% in the original population) exhibit marked excitation with convulsive fits in response to the action of

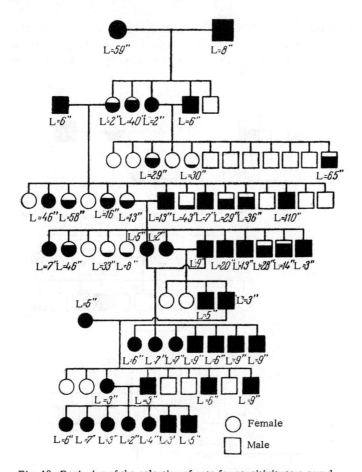

Fig. 18. Beginning of the selection of rats for sensitivity to a sound stimulus. Solid squares (or circles) denote rats showing a pathological excitation in 100% of tests; shading illustrates the proportion of tests in which a pathological excitation appears in response to the action of the sound stimulus; outlined symbols denote rats not exhibiting pathological excitation during the action of the sound stimulus. L is the mean duration of the latent period of the motor excitation reaction.

the sound stimulus. A large proportion of the rats of this line gives a single-wave excitation reaction with a short latent period, which as a rule is more intensive and culminates more frequently in a convulsive fit than does the double-wave type.

The creation of a line of rats highly sensitive to the sound stimulus enabled us to carry out many physiological and patho-physiological investigations.

The selection of animals showing low sensitivity to the sound stimulus was also effective. The reaction of rats of this line was characterized by a long latent period and it rarely culminated in a convulsive fit (Fig. 19).

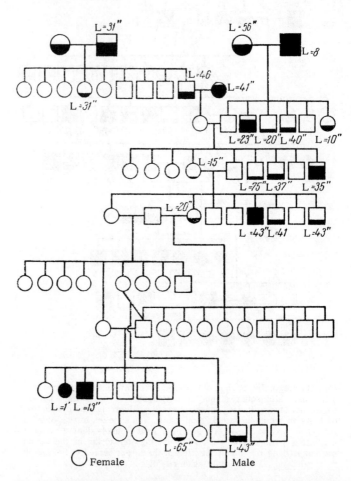

Fig. 19. Beginning of the selection of rats for insensitivity to a sound stimulus. Solid squares (or circles) denote rats showing a pathological excitation in 100% of tests; shading illustrates the proportion of tests in which a pathological excitation appears in response to the action of the sound stimulus; outlined symbols denote rats not exhibiting pathological excitation during the action of the sound stimulus. L is the mean duration of the latent period of the motor excitation reaction.

The efficacy of selection when directed both toward increase or decrease of the sensitivity of rats to the sound stimulus demonstrated the innate nature of this property of the nervous system (Krushinskii, 1959). When we speak of the innate nature of this reaction, however, it must not be forgotten that certain external conditions may change considerably the sensitivity of rats to the action of a sound stimulus.

A deficiency of magnesium in the diet (Kruse et al., 1932; Greenberg et al., 1938) or of vitamin B_1 (Patton et al., 1942, 1942a), and many other external factors may considerably increase the sensitivity of rats to a sound stimulus. Rats not reacting to this stimulus, if kept on a diet deficient in the above substances began to react with excitation and convulsions.

By the selection of rats highly sensitive to the action of the sound stimulus, we created a line of highly excitable animals with a weak process of inhibition. As our investigations showed, a short latent period and an intensive convulsive fit are characteristic of high excitability; a long first wave of excitation and a short period of inhibition between the two waves of excitation are characteristic of weakness of the process of inhibition. Single-wave excitation reactions are characteristic primarily of a relatively weak state of inhibition, and also of the high excitability of the nervous system. Members of our highly sensitive line, by comparison with those of the line of low sensitivity or with rats taken from a random population, are, in fact, characterized by these signs.

By the creation of a line of rats with a constant pathological excitation reaction, terminating in convulsive fits, we thus, as a result of selection, increased the excitability of their nervous system and weakened their process of inhibition, for the ratio between these properties of the nervous system, as we shall show below, determines the manifestation and the character of expression of this pathological state.

No unanimity exists regarding the physiological mechanisms lying at the basis of the reaction of rats to a sound stimulus. Some writers have used the term "neurosis" to describe the reaction of rats and mice to the stimulus used. The convulsive phenomena were regarded as neurotic, arising as a result of a "conflict" between different manifestations of nervous activity.

Vasil'ev (1924), for instance, who made the first attempt to explain the physiological nature of the convulsive fits observed by Studentsov and Ganike in mice, regarded this phenomenon as a collapse of nervous activity. The conflict between the excitation caused by presentation of the feeding bowl, and the inhibition arising, in his opinion, from the strong sound stimulus, produces a state of strong excitation in the animal, which may terminate in convulsive fits.

In 1939, Maier (cited by Gentry and Dunlap) attempted to explain the fits by a "conflict" between the different manifestations of nervous activity. In order to produce convulsive fits he used a strong sound stimulus, impelling the rat to jump onto a platform which it had previously been taught to avoid. This situation, in Maier's opinion, led to an excessive strain on the nervous processes, culminating in a convulsive fit, which he called "neurotic behavior."

The experiments of Maier and others made it clear, however, that in order to cause excitation followed by convulsive fits in both rats and mice it was sufficient to give the sound stimulus alone, and that, on the contrary, a "conflict" situation without the sound stimulus did not lead to convulsive fits (Finger, 1945). Another supporter of the "conflict theory," Bitterman (1944), considered that "all situations in which abnormal behavior of an animal is observed may be regarded as situations of conflict," and the convulsive fit in rats in response to a sound stimulus is the result of a "conflict" between the desire to avoid the strong sound and the impossibility of doing so because of the confined space in which the rat is placed during its exposure to sound. This hypothesis too, however, has not been confirmed experimentally. A convulsive fit in response to a sound stimulus may be obtained in rats which are not shut up in the confined space of an experimental chamber or cage, and are able to get away from the action of the sound.

"Conflict" is thus not the essential cause of the development of motor excitation and the convulsive fit, and the "conflict theory" does not provide the correct physiological explanation of this phenomenon.

We considered that the physiological mechanisms lying at the basis of the excitation of rats and mice in response to a sound stimulus may be explained by an initial perspective that the basis of the marked excitation ending in convulsive fits is general disturbances of the processes of excitation and inhibition.

The starting point of our physiological analysis was the fact, described above, that the excitation in rats in response to the action of a sound stimulus may be manifested in two forms. We considered that the development of excitation in waves (in response to the action of a sound stimulus) is the outward reflection of the physiological changes responsible for the manifestation of this pathological state.

Arising from Pavlov's concept of the two fundamental processes lying at the basis of higher nervous activity—excitation and inhibition—we postulated that the existence of waves during the development of the fit is due to the "conflict" between these processes. According to our view (Krushinskii, 1949; Krushinskii, Fless, and Molodkina, 1950), the overloading of the auditory analyzer of the rat by the action of a continuous sound stimulus leads to wide radiation of the process of stimulation along the neurons of the brain, and this radiation, affecting the motor centers, leads to the manifestation of motor activity by the animal. In response to the initial excitation, inhibition is induced, which is a physiological measure of "conflict" with the beginning pathological excitation. The appearance of a period of inhibition after the first outburst of excitation is the result, in our opinion, of induced relationships between the processes of stimulation and inhibition. The continuing action of the sound stimulus leads to a considerable strain on the process of inhibition, and to its subsequent exhaustion.

Pavlov (1938) stated that "by increasing the state of inhibition in the cell not gradually, but suddenly, by the action of a suitable external stimulus, we cause an extraordinary weakening of the inhibitory function of the cell, almost abolishing it."*

After the exhaustion or abolition of the process of inhibition, as a result of the continuing action of the sound stimulus, the excitation develops afresh. This appears in the form of the second wave of motor excitation of the animal. The process of stimulation, unrestrained by the more exhausted inhibition, radiates widely along the neurons of the brain, as a result of which a convulsive fit may arise. In those animals in which the process of excitation has a tendency toward rapid radiation and the process of inhibition is relatively weak, so that it cannot inhibit the beginning excitation of the brain, only one wave of motor excitation develops, without an inhibitory phase.

Thus, according to our working hypothesis, the pathological excitation developing in rats in response to the action of a sound stimulus is based upon the quantitative relationship between the two fundamental processes of nervous activity: excitation and inhibition. Their mutual relationship determines the character of the motor excitation developing upon exposure of the animal to sound (Fig. 20).

*I. P. Pavlov, op. cit., p. 684.

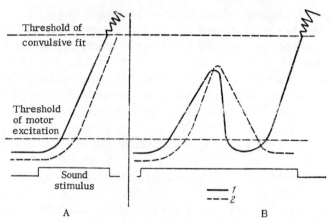

Fig. 20. Diagram showing the mutual relationship of the processes of excitation and inhibition in response to the action of a sound stimulus: A—single wave reaction; B—double wave reaction. 1—Process of excitation; 2—process of inhibition.

The functional intensification of the process of inhibition and the decrease in the degree of excitability of the nervous system must lead to the appearance of a double wave of motor excitation and the weakening of the intensity of the convulsive fit, and conversely a weakening of the inhibitory function of the nerve cell and an increase in the excitability of the nervous system must bring about the development of a single wave of excitation and the intensification of the convulsive fit.

We used bromide as the drug to strengthen inhibition. Our experiments showed that the administration of sodium bromide (from 40 to 60 mg, 30-40 minutes before the start of the experiment) to rats with a single-wave reaction of excitation leads to the development of two waves of excitation, with a period of inhibition between them (Fig. 21). The intensity of the convulsive fit weakens with an increase in the dose of bromide. Bromide diminishes the intensity of the first wave of excitation, which, as the dose of bromide is increased, is shortened to 1-2 seconds or does not even appear. The strengthening inhibition process apparently "extinguishes" it. However, the inhibition, which is exhausted during the inhibitory period, does not change the character of the second wave of excitation. A further increase in the dose of bromide has the result that the rats generally cease to react by motor excitation to the sound stimulus.

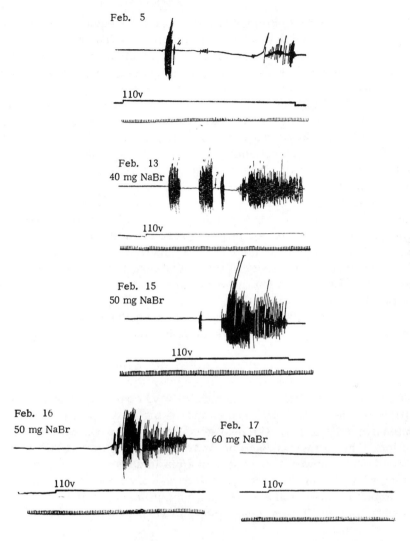

Fig. 21. Change in the character of manifestation of motor excitation in the same rat under the influence of different doses of bromide.

Administration of bromide leads not only to the development of an inhibitory phase between two waves of excitation, but also to the prolongation of this phase. Experiments conducted by Fless on 19 rats which, before receiving sodium bromide, as a rule exhibited one wave of motor excitation, showed that the period of time between the waves of excitation developing in the rats after receiving specific doses of bromide (individually determined for each rat) increased considerably as the dose of bromide increased. The period of inhibition for minimal doses of bromide (at which an inhibitory period first appeared in the rats) was 15.7 ± 1.47 seconds. By increasing this dose of bromide this period could be lengthened to 24.4 ± 2.61 seconds. The difference of 8.7 ± 2.99 seconds, being 2.9 times greater than the probable error, is statistically significant (Fless, 1957).

There is no doubt that the lengthening of the period of inhibition is the result of strengthening of the process of inhibition, which withstands the strain for longer and is not exhausted by the action of the sound stimulus.

Experiments in which bromide was administered thus showed that a strengthening of the inhibitory process leads to a consistent change in the character of the motor reaction in the rat: the appearance of a double wave instead of a single wave of excitation, the weakening of the intensity of the convulsion, the lengthening of the inhibitory period between the waves of excitation, and the shortening of the first wave of excitation.

Complete agreement with the findings described above is shown by the results of experiments in which the process of inhibition was weakened.

These investigations (Krushinskii, Molodkina, and Kitsovskaya, 1950) showed that extirpation of the parathyroid glands in rats in which the excitation is of the double-wave character leads to abolition of the inhibitory phase and to the development of a more intensive convulsion, which always follows the first wave of excitation. The investigations of Andreev and Pugsley (1935) and of Bam (1939) showed that hypercalcemia in dogs causes a strengthening of inhibition, so that hypocalcemia must evidently be regarded as a factor essentially weakening inhibition, which in normal rats resists the excitation developing in response to the action of a strong sound stimulus.

The blood calcium level in normal (nonparathyroidectomized) rats, however, does not determine the sensitivity of the rat to the action of a sound stimulus. Kitsovskaya's experiments showed no positive correlation between the calcium content of the serum and the relative sensitivity of rats to a sound stimulus (Table 26).

The results in Table 26 show that: 1) the difference between the rats sensitive or insensitive to the sound stimulus is not determined by the calcium content of the blood serum; 2) rats previously insensitive to the sound stimulus after parathyroidectomy become sensitive or continue to remain insensitive, also irrespective of the degree of lowering of the blood calcium concentration.

Table 26. Calcium Content of the Blood Serum in Normal and
Parathyroidectomized Rats Sensitive and Insensitive to
a Sound Stimulus

Group of animals	Insensitive rats			Sensitive rats		
	Number of rats	Mean Ca content (in mg %)	Amplitude of variations (in mg %)	Number of rats	Mean Ca content (in mg %)	Amplitude of variations (in mg %)
Normal rats .	19	10.61	9.8–11.9	23	10.54	9.5–12
Parathyroid-dectomized rats (before operation, "insensitive")	10	7.3	4.5– 8.0	8	7.4	6.5– 8.0

An increase in the blood calcium concentration in parathyroid-ectomized rats (1 ml of a 10% $CaCl_2$ solution was injected every 3 hours), however, led to a considerable weakening in the intensity of the epileptiform reaction (Table 27).

Table 27. Action of Calcium Chloride on the Excitation and Reflex
Epileptiform Fit in Parathyroidectomized Rats

Number of rats	Before injection of $CaCl_2$		During injection of $CaCl_2$		After administration of $CaCl_2$ ceased
	Blood calcium (in mg %)	Mean intensity of reaction (points)	Blood calcium (in mg %)	Mean intensity of reaction (points)	Mean intensity of reaction (points)
5	7.0	2.6	9.6	0.6	1.8

It may be seen from Table 27 that as a result of the injection of calcium into parathyroidectomized rats, parallel to the increase in the blood calcium concentration there is a considerable weakening of the intensity of the reaction (in two rats receiving calcium injections the sound stimulus generally did not cause excitation). After administration of calcium had ceased, the intensity of the reaction again increased. These results show that a decrease in the calcium content of the blood serum lowers the threshold of sensitivity of rats to the sound stimulus.

By the administration of bromide to parathyroidectomized rats,
however, it is possible to so strengthen inhibition that, in spite of
the low level of blood calcium, the animals cease to respond by
excitation to the sound stimulus (Table 28).

Table 28. Results of Administration of Sodium Bromide to
Parathyroidectomized Rats (50 mg daily for 5-6 days)

Number of rats	Before administration		After administration	
	Mean intensity of reaction (points)	Mean blood calcium content (in mg %)	Mean intensity of reaction (points)	Mean blood calcium content (in mg %)
9	3.3	6.5 limits of variation 3.2–8.1	0	6.6 limits of variation 2.7–8.0

It will be seen from Table 28 that bromide, without changing
the blood calcium level, completely abolishes the reaction of excitation of the animal in response to the sound stimulus.

The administration of smaller doses of bromide (10-35 mg
daily) or of calcium chloride (0.25–0.50 cc of a 10% solution six
times a day) to parathyroidectomized rats (n = 18) led to the regular
appearance of an inhibitory phase in the reaction of excitation,
which thus reverted to the preoperation double-wave type.

The results of these experiments show that marked excitation
and convulsive fits in rats in response to the action of a sound
stimulus after parathyroidectomy are the result of weakening of
the process of inhibition, in consequence of the hypocalcemia caused
by removal of the parathyroid glands.

Experiments in which the excitability of the nervous system
was altered were also in full agreement with our working hypothesis
regarding the importance of the mutual relationship of stimulation
and inhibition in determining the character of motor excitation and
the intensity of the convulsive fit.

As one of the methods of weakening of the developing excitation,
we reduced the intensity of our sound stimulus (Krushinskii, 1954).
A decrease in the intensity of the sound from 112 to 70 decibels
was achieved by reducing the voltage of the current supplying the
bell (this change, as described above, was expressed conventionally
in volts—from 130 to 20 v). The primary effect of the decrease in
intensity of the sound was a considerable lengthening of the latent
period.

The problem of the change in the latent period in relation to the intensity of the sound stimulus was discussed in the paper by Morgan (1941). As a stimulus to produce the reaction in rats this worker used a Galton's whistle. Morgan found that when the strength of the stimulus was changed the latent period, i.e., the time from the beginning of application of the sound stimulus to the beginning of the motor excitation of the animal, did not alter. He accordingly concluded that the nervous processes responsible for the epileptiform reaction obey the "all or nothing" principle.

Investigations carried out in our laboratory by Fless and Vasil'eva, and also the experiments of Malinovskii (1954) did not confirm Morgan's conclusions. It was found that a change in the strength of the sound stimulus alters the duration of the latent period of the reaction in the animal. Of the 45 rats taking part in the experiment, in 42 (95.5%) the latent period was lengthened as the intensity of the sound stimulus was weakened. Only in two rats was a tendency observed for the latent period to be shortened as the sound was weakened, and in one animal the latent period was unchanged.

The influence of the intensity of the sound stimumus on the duration of the latent period may also be clearly seen by comparing the mean values of the latent periods in the same group of rats (n = 22) exposed to the action of a bell, the loudness of which was gradually decreased from one experiment to the next. When the bell was acting at maximum intensity (current 120v) the latent period averaged 5.36 ± 0.56 seconds, when it was acting at moderate intensity (current 60-80v) the mean value was 11.82 ± 1.41 seconds, and when the bell was at the minimum intensity at which the rats still exhibited an excitation reaction (current 20-40v), the duration of the latent period was 22.0 ± 1.97 seconds. The difference between the three indices was statistically significant in all cases.

On the basis of these experiments it is thus clear that the duration of the latent period is directly dependent on the intensity of the sound stimulus.

The physiological processes lying at the basis of the excitation arising under the influence of a sound stimulus thus obey the law of strength relationships as established by the work of Pavlov's school for conditioned reflex activity, and not the "all or nothing" law.

With a change in the strength of the sound stimulus, the number of waves of motor activity of the animal also undergoes a regular change.

Obvious changes take place in those individuals in which the motor excitation preceding the convulsive fit develops in the form of a single wave. A decrease in the intensity of the sound as a rule leads to the appearance of two waves of excitation, separated by a period of inhibition; with an even greater weakening of the sound the rat generally ceases to exhibit an excitation reaction (Fig. 22).

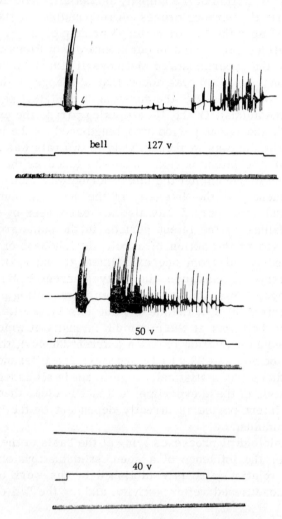

bell   127 v

50 v

40 v

Fig. 22. The effect of weakening of the intensity of the sound stimulus on the character of the motor excitation in the rat. The number is inserted at the occurrence of the fit, and denotes its intensity.

For instance, of 22 rats giving one wave of excitation in response to a strong sound stimulus, 19 began to react with two waves of excitation when the strength of the sound stimulus was reduced. Only three rats continued to react with one wave of excitation in spite of the considerable (limiting) weakening of the sound of the bell by means of a decrease in the current to 20-30v (70-80 db) (experiments of Fless). It is evident that a still greater decrease in the strength of the sound would also have led in these rats to the development of a double-wave excitation reaction.

A weakening of the intensity of the stimulus applied, leading to weaker excitation of the neurons of the brain, thus brings about the development of an inhibitory phase, and the reaction acquires the character of a double wave.

In rats with two waves of excitation a weakening of the intensity of the sound stimulus leads to lengthening of the period of inhibition between the first and second waves of excitation. As Molodkina's experiments showed, the period of inhibition between the waves of excitation, which is 14.5 ± 0.8 seconds with a stronger sound stimulus (115-80v supplied to the bell), is prolonged to 18.67 ± 1.19 seconds when the sound stimulus is weakened (70-40v supplied to the bell). The difference of 4.17 ± 1.48 seconds, which is 2.75 times the probable error, is practically significant. A weaker sound stimulus, leading to a less marked excitation of the neurons of the brain, takes longer to exhaust the restraining process of inhibition; this is manifested outwardly by a lengthening of the period of inhibition between the two waves of excitation.

Parallel with the lengthening of the inhibitory phase, the first wave of excitation is shortened. When the sound stimulus is weakened in intensity, the duration of the first wave of excitation is slightly diminished (from 9.58 ± 0.69 seconds to 8.00 ± 0.36 seconds). The difference of 1.58 ± 0.63 seconds (significance of difference 2.51) approximates to the limits of statistical significance. There is no doubt that the weaker excitation caused by the less intensive sound stimulus is more quickly suppressed by the inhibitory process than the stronger excitation caused by the stronger stimulus.

Finally, a weakening of the intensity of the sound stimulus considerably modifies the proportion of cases of excitation terminating in convulsive fits.

Experiments conducted by Fless showed that as the intensity of the sound was diminished from experiment to experiment the convulsive fits became less intensive in the majority of rats (12 of 22). When the intensity of the sound was further reduced, in 19 of the 22 rats the convulsive fits disappeared; the reaction of the rats was expressed only in the form of motor excitation. When the intensity of the sound of the bell was very low (individually determined for each rat), no motor excitation was exhibited by 20 animals.

Experiments in our laboratory in which we administered drugs which increase the excitability of the nervous system (caffeine, strychnine) also confirmed the hypothesis of the dynamic inter-relationship between the processes of excitation and inhibition, on which we based our study of this particular pathological state. Obvious changes are observed, especially in the length of the latent period. When the rats were given caffeine (caffeine sodium benzoate) in doses of 50-70 mg/100 g body weight, the latent period was shortened from 15.8 ± 1.42 seconds to 7.6 ± 0.67 seconds. The difference of 8.2 ± 1.57 seconds is statistically significant.

When discussing the action of caffeine on the duration of the latent period, mention must be made of the role of the individual sensitivity of rats to this drug. Cases have been observed in which the latent period was lengthened as the dose of caffeine administered was increased. In rat No. 2755, for instance, when the dose of caffeine was increased, the latent period was lengthened. When caffeine was given in a dose of 15 mg/100 g body weight the latent period was 6 seconds, a dose of 20 mg—7 seconds, 25 mg—8 seconds, 50 mg—10, and 70 mg—13 seconds.

It may be suggested that in such cases the working capacity of the nerve cells of the rats is approaching its limit, as is observed in the conditioned reflex activity of dogs with a weak type of nervous system when they are given large doses of caffeine.

The administration of caffeine changes the character of the waves of excitation preceding the convulsive fit. Of 15 rats which systematically gave two waves of excitation in response to the action of the sound stimulus, after administration of caffeine a single wave of excitation appeared in eleven. The development of a single wave of excitation depends on the dose of caffeine given. For instance, when caffeine was given in doses of 5, 10, and 15 mg per 100 g body weight, a single wave of excitation was observed in 29.5% of cases (in 13 of the 44 tests); when the dose was 35, 50, or 70 mg/100 g body weight, in the same rats the number of cases showing a single wave of excitation was nearly doubled, reaching 55.5% (25 of 45 tests).

These results demonstrate that when the excitability of the nervous system of rats is increased with caffeine, two waves of excitation are replaced by one.

Despite the obvious changes in the duration of the latent period and in the character of the waves of excitation caused by administration of caffeine, we were unable to discover any essential change in the duration of the first wave of excitation in the rats with a double wave of excitation. Before the administration of caffeine the mean duration of the first wave of excitation was 9.2 ± 0.51 seconds (n = 30 tests); after the administration of caffeine the value was 10.3 ± 0.26 seconds. This small increase in the duration of the first wave of excitation is not statistically significant: the difference of 1.1 ± 0.97 seconds hardly exceeds the probable error.

After the administration of caffeine, changes are observed in the duration of the period of inhibition between the first and second waves of excitation. The mean duration of the period of inhibition in the experimental group of rats before receiving caffeine was 20.2 ± 2.04 seconds (29 trials); after administration of caffeine in all the doses used the duration of inhibition was shortened to 13.6 ± 0.73 seconds (61 trials). The difference between the two values is statistically significant, amounting to 6.6 ± 2.2 seconds, i.e., three times the probable error. The experiment shows that the process of inhibition is more quickly exhausted if it suppresses a stronger excitation of the neurons of the brain, caused in this case by the action of the sound stimulus in association with caffeine.

After administration of caffeine regular changes were observed in the duration of the second wave of excitation. In all the experimental rats (in all 30 trials) during the control period (i.e., before the beginning of administration of caffeine) the second wave of excitation, if it did not terminate in a convulsive fit, ended abruptly at the moment the sound stimulus was removed. When the rats were exposed to sound after the preliminary administration of caffeine, in almost half the cases (25 of 53) the second wave of excitation did not end immediately after the sound stimulus was removed, but continued a short time longer. The minimum duration of this prolongation of excitation (the "inertia of the process of stimulation") was 15 seconds, and the maximum was 18 minutes.

In most cases, when a single wave of excitation without a convulsive fit developed in rats under the influence of caffeine, this wave also continued after the action of the sound stimulus had ceased. For instance, in 15 of 24 cases of single-wave excitation without convulsion the wave did not come to an end after removal of the sound stimulus. In some cases it lasted over 10 minutes.

These facts are evidence that when the excitability of the nervous system is increased by means of caffeine, a sound stimulus causes marked stationary changes in the state of excitability of the nervous system.

The administration of caffeine increases the intensity of the convulsive fit. As Molodkina showed by her investigations, the mean intensity of the convulsive fit, which in the group of experimental rats (n = 22) was 2.6 ± 0.18 points, rose after the administration of caffeine to 3.6 ± 0.15 points. The difference of 1.0 ± 0.23 points is statistically significant.

Experiments with strychnine (doses of 25–175 $\mu g/100$ g body weight, by subcutaneous injection) (Krushinskii, Fless, and Molodkina, 1950) gave similar results to the experiments in which caffeine was given.

In the first place strychnine, like caffeine, shortens the latent period of the excitation reaction. For instance, the mean duration of the latent period in the group of experimental rats (n = 13) was shortened from 18.43 ± 4.94 seconds (before injection of strychnine) to 9.36 ± 5.21 seconds (after injection). Although the difference is not statistically significant (because of the greater variation in the duration of the latent period in the group of rats being studied), it is nevertheless great enough to demand consideration.

Second, there was an increase in the number of cases with single-wave motor excitation. Of the 14 trials of 13 rats sensitive to the sound stimulus, in only two cases (14.3%) was a single wave of excitation observed; after injection of strychnine seven trials terminated in single-wave excitation (50%).

Third, strychnine increases the intensity of the convulsive fit. Of the nine cases where convulsive fits terminated excitation reactions in the group of rats which we studied, there was no fit of maximal, tonic degree (4 points); a clonic convulsive reaction was observed in nine cases. After the injection of strychnine, of ten convulsive fits nine were tonic and only one was clonic.

These experiments in which the excitability of the nervous system was modified showed that the individual components of the motor excitation and the convulsive fit characterize the state of excitation and inhibition of the nervous system.

1) The duration of the latent period of development of the excitation reaction characterizes the excitability of the animal's nervous system: the shorter the latent period the higher the excitability of the neurons of the brain concerned in the pathological reaction.

2) Single-wave excitation characterizes both the increased excitability of the nervous system and, in particular, the relative weakness of the process of inhibition.

3) Double-wave excitation is evidence of a relatively low level of excitability of the nervous system, and especially of the considerable strength of inhibition.

4) The short (abortive) first wave of excitation is evidence in particular of a relatively strong process of inhibition.

5) The lengthening of the period of inhibition between the first and second waves of excitation denotes both a decrease in the degree of excitability of the nervous system and a strengthening of the inhibitory function of the nervous system.

6) The prolonged inertia of excitation after the second wave of excitation is evidence of the high excitability of the nervous system.

7) An increase in the intensity of the convulsive fit is evidence both of an increased excitability and of a weakening of the inhibitory function of the nervous system.

This model may be found useful in the pharmacological evaluation of new drugs. It makes possible the rapid evaluation of their specific action on the fundamental processes which are the basis of the activity of the nervous system.

We carried out a series of investigations into the role of inhibition in countering the state of excitation of the brain. The experiments showed that inhibition, which actively restrains the excitation induced by a sound stimulus, is evidently exhausted, so that the second wave of excitation develops against the background of the previously exhausted process of inhibition.

When experiments were carried out on rats with a double-wave excitation reaction, the sound stimulus was removed at the moment of maximum development of each wave of excitation. It was found that the inertia of the first wave of excitation, developing before the process of inhibition was exhausted, continued on the average for a further five minutes, whereas the inertia of the second wave of excitation, which developed after the process of inhibition had become exhausted, was 104.9 seconds in duration, i.e., 20 times that of the first wave of excitation.

Increasing the excitability of the nervous system of the rat by the preliminary administration of caffeine does not alter the duration of the inertia of the first wave of excitation, but it does considerably lengthen the inertia of the second wave of excitation (Fig. 23). These experiments clearly showed that the process of inhibition, which is used successfully by the nervous system as a

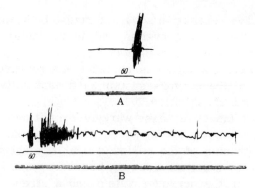

Fig. 23. Inertia of the first and second waves of
excitation (after administration of caffeine): A—
when the sound stimulus was removed at the
maximum of the first wave of excitation; B—at
the maximum of the second wave of excitation.

defensive mechanism against the development of a sudden and
marked excitation, becomes exhausted during the continuing action
of the stimulus, as is revealed by the subsequent prolonged state
of motor excitation, and sometimes by a convulsive epileptiform fit.

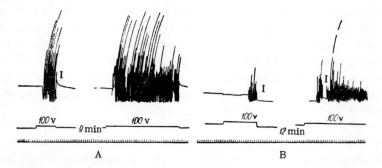

Fig. 24. Nonrecovery of the process of inhibition after 9 minutes (A) and
its commencing recovery after 12 minutes (B); I—inhibitory phase of fit.

Having established the fact that the process of active inhibition
may become exhausted during the action of the sound stimulus, we
were then able to study the time and the conditions of its subsequent
recovery. Experiments were conducted as follows. Using rats
which systematically gave a double-wave reaction of excitation in
response to the action of the sound stimulus, we switched off the
bell during the inhibitory phase and then switched it on again after
various intervals of time. One wave of excitation in response to the
second ringing indicated that the process of inhibition in the rat

had not yet been restored (Fig. 24A). Two waves of excitation with a short inhibitory phase between them indicated the beginning of restoration of the exhausted process of inhibition (Fig. 24B).

By increasing the length of the interval between the two ringings from one experiment to the next, we were able to determine accurately the minimum time of appearance of two waves of excitation, i.e., the time of recovery of the process of inhibition. This varied in different rats (n = 216) from 15 seconds to 20 minutes. Administration of bromide shortens the time for restoration of the inhibition process. For instance, in a group of rats with a mean time of recovery of the inhibition process of 16.7 minutes (limits of variation 11–24 minutes), the mean recovery time was shortened to 10.8 minutes (limits of variation 7–16 minutes).

The foregoing remarks show the importance of the process of inhibition in the pathological excitation reaction developing in rats in response to the action of a sound stimulus.

These investigations thus show that upon marked excitation of the neurons of the brain, a defensive role is undertaken by the process of inhibition induced by this excitation, and when this is of adequate strength it arrests the commencing excitation, although this inhibition is unstable and is easily exhausted during prolonged excitation of the brain.

The nature of the physiological characteristics of this type of inhibition presents a difficult problem. Without having sufficient grounds for describing it by some new term, we consider that in its properties this type of inhibition bears the closest resemblance to extinctive inhibition, although it arrests the excitation caused by an unconditioned reflex stimulus. Should two waves of excitation develop, evidently what is happening is the acute (although temporary) extinction of the excitation which develops in response to the action of the specially applied sound stimulus.

Pavlov pointed out that active inhibition manifests itself in the behavior of an animal outside its conditioned reflex experience. "Its relaxation leads to the abnormal predominance of the process of stimulation in the form of the disturbance of differentiation, of delay, and of other abnormal phenomena in which inhibition plays a part, and it is also shown in the animal's general behavior in the form of irritability, impatience, and rage, and finally, in the form of pathological phenomena."*

The most probable explanation is that active (especially extinctive) inhibition will suppress any form of excitation of the brain,

*I. P. Pavlov, op. cit., p. 735.

whether of conditioned or of unconditioned reflex origin. We therefore assume that the inhibition which is exhibited by rats during
their motor "rage" is active, but is evidently not a cortical inhibitory process.*

The question arises which defensive mechanisms for the suppression of the excitation caused by the sound stimulus are used
by the nervous system after the exhaustion of this form of inhibition.
Investigations carried out in our laboratory showed that under
these circumstances a parabiotic, limiting inhibition is used. After
the active inhibition is exhausted and the sound stimulus continues
to act, the reaction of the animal's nervous system to the stimulus
applied displays several distinctive features, irrespective of whether
or not a convulsive fit took place (Krushinskii, Fless, and Molodkina,
1952). The main feature characterizing the state of the nervous
system at this period is its extraordinarily increased excitability.

This increased excitability shows itself, in the first place, by
the fact that the rat reacts with motor excitation (in response to
a sound stimulus) with an extremely short latent period (instead
of one of 5–10 seconds, one of tenths or even hundredths of a second). Only after an interval measured in tens of minutes is the
excitability of the animal's nervous system restored.

The latent period of the motor reaction of the animal was measured by means of graphic recording on a rapidly revolving kymograph (880 mm/sec). As a time marker we used a tuning fork giving
100 vibrations per second. The recording of the beginning of the
animal's reaction and of the moment of application of the stimulus
enabled us to measure the latent period of the animal's excitation
reaction.

In order to reveal the degree of excitability and the phases of
limiting inhibition after the action of the sound stimulus for 1.5
minutes, the rat was exposed for a few minutes to the interrupted
action of a sound stimulus (alternation of strong and weak sound
stimuli, each lasting 10 seconds, separated by intervals of 10
seconds).

The increased excitability after the action of the sound stimulus
is also indicated by the lowering of the threshold of the reaction of
the rats to this stimulus. Of 29 rats tested after excitation caused
by a strong sound stimulus, for instance, 25 (86%) now began also
to react by excitation to a weak sound, which had previously been
subthreshold (an 80 db bell).

*The cortical nature of this inhibition is opposed by the investigations of Kotlyar, who
showed in our laboratory that excitation may be interrupted by an inhibitory period.

During the prolonged action of a sound stimulus, a further increase in excitability is observed, which takes the form of prolonged inertia of the motor excitation after cessation of the action of the stimulus.

The considerable shortening of the latent period, the lowering of the threshold of reaction of the rats to the sound stimulus, and, finally, the state of continued excitation after the prolonged action of the sound stimulus are all evidence of a marked increase in the functional activity of the neurons of the brain.

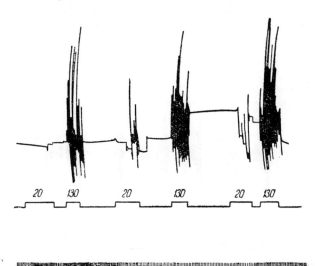

Fig. 25. Correct strength relationships.

We consider that this fact is of great importance for the pathophysiology of nervous activity. It points to the role of nervous trauma as a factor causing a stationary increase in the excitability of the nerve centers—a phenomenon which is evidently at the root of most cases of abnormal nervous activity. This condition is reminiscent of the phenomenon of increased excitability of the central nervous system after prolonged stimulation of a sensory nerve, discovered by Vvedenskii (1912). He saw the similarity between this phenomenon and the symptom-complex of hysterical manifestations in man, and he therefore called the increased excitability of the nervous system taking place as a result of its prolonged excitation by the name "hysteriosis."

Against the background of increasing excitability of the nervous system in rats, phases of limiting inhibition begin to appear, and moreover, particular phases of limiting inhibition correspond to particular degrees of excitability (Krushinskii, Fless, and Molodkina, 1952).

The most convenient way of revealing the phases of limiting inhibition, as we have pointed out above, is to use short sound stimuli (10 seconds) of different intensity, at intervals of 10 seconds. By this method we were able to observe stages of parabiotic, limiting inhibition, discovered by Vvedenskii and Pavlov, and also to find new relationships between stimulation and the response reaction of the nervous system.

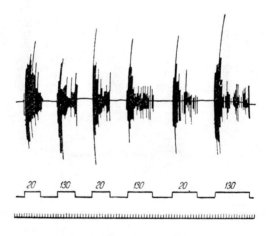

Fig. 26. Balanced stage.

When the degree of excitability of the nervous system of an animal is relatively low, the latent period can be measured in seconds. Strength relationships are observed between the intensity of the sound and the degree of the animal's response reaction: in response to a strong sound stimulus the animal reacts with intensive motor excitation with a relatively short latent period; to a weak sound stimulus the animal either does not react at all, or it reacts with weak motor excitation with a longer latent period (Fig. 25). In these cases the limiting inhibition has not yet distorted the normal relationship between the intensity of the stimulus and the response reaction of the animal.

With higher degrees of excitability of the nervous system, the normal strength relationships become replaced by a balanced stage in which the animal exhibits the same intensity of motor excitation in response to strong and weak sounds (Fig. 26). The balanced stages proceed at a low and a high level. Balanced stages often give way to paradoxical stages when, under the action of a strong sound stimulus, the rat develops a weaker motor reaction than in response to a weak stimulus (Fig. 27).

The change from a strength relationship to a balanced and, finally, to a paradoxical stage takes place against a background of gradually increasing excitation of the higher divisions of the animal's nervous system. This may be seen clearly from the results of measurement of the latent period at various stages of parabiotic inhibition (Krushinskii, Fless, and Dubrovinskaya, 1957) (Table 29).

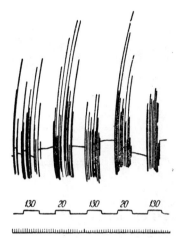

Fig. 27. Paradoxical stage.

It is clear from Table 29 that each stage is characterized by a definite magnitude of the latent period, which becomes shorter as the parabiotic state becomes more advanced. The highest excitability is characteristic of the paradoxical stage.

Table 29. Duration of Latent Periods in Various Stages
of Parabiosis

Intensity of bell	Mean latent period (in seconds)			
	Strength relationships	Balanced at a high level	Balanced at a low level	Paradoxical
Weak	1.56 ± 0.6	0.87 ± 0.2	0.55 ± 0.12	0.30 ± 0.08
Strong.............	1.18 ± 0.12	0.51 ± 0.07	0.33 ± 0.05	0.21 ± 0.07

These findings show that as the parabiotic state becomes more advanced, the excitability of the central nervous system is increased.

The reactions of the experimental rats in these phases sometimes have an explosive character: in response to stimulation the animal at once develops a sudden excitation, which then rapidly weakens (Fig. 28). A like phenomenon, observed in the conditioned

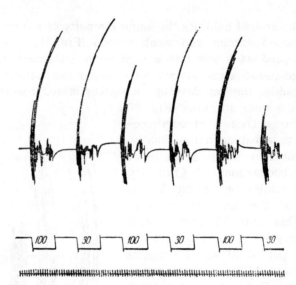

Fig. 28. Explosiveness.

reflex activity of dogs, was termed explosiveness or stimulatory
weakness by Pavlov. Its presence indicates the development of
limiting inhibition in these conditions, which arises immediately
after the marked excitation.

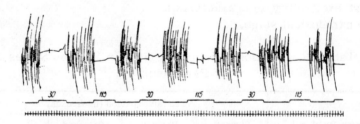

Fig. 29. Inverted stage.

With a further increase in the excitability by the interrupted
action of the sound stimulus the rat begins to exhibit motor excitation not only in response to the stimuli used, but also to the intervals
between them. The functional state of the nervous system during
continued excitation is evidently similar to that which Pavlov described as "pathological inertia of the process of stimulation."
    The application of a sound stimulus against the background of
continued excitation cannot alter the course of the excitation. In
most cases, however, the application of a sound stimulus weakens

or completely abolishes the excitation, and when the bell is switched off the excitation begins again. A complete reversal of the rat's reaction to the sound stimulus takes place (removal of the stimulus causes excitation and application of the stimulus causes the cessation of excitation) (Fig. 29). It is evident that the marked increase in the excitability of the nervous system, manifesting itself as the prolongation of excitation, is restricted during the further application of the traumatizing stimulus by limiting inhibition, which outwardly takes the form of the cessation of the motor activity of the animal.

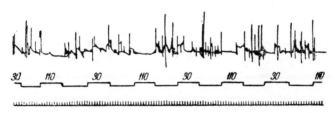

Fig. 30. Inverted stage. Inhibiting-strength relationships.

The regular appearance of these distinctive relationships between the action of the stimulus and the response reaction of the animal enables us to regard this state of the nervous system as a definite stage in the development of limiting inhibition. It bears some external resemblance to Pavlov's "ultraparadoxical phase" and to Vvedenskii's "inhibiting stage," although it is not identical with either.

In the ultraparadoxical phase the animal reacts with a response reaction to an inhibitory (differential) stimulus, but does not react to a positive conditioned stimulus.

During an inhibiting stage strong stimuli, coming from a normal part of the nerve, inhibit the effect of another stimulus acting at the same time on the parabiotic area.

In our experiments the animal displayed motor excitation in the absence of stimulation, and inhibition of this excitation during the action of the sound stimulus. We therefore called this stage an "inverted stage" (Krushinskii, Fless, and Molodkina, 1952).

The inhibiting effect during the action of the stimulus, which is characteristic of this stage, may be regarded as the consequence of the summation of two excitations: the existing stationary excitation of the central nervous system, caused by preceding stimuli, and the excitation from the action of the sound stimulus, applied against the background of the stationary excitation. As a result of this summation the excitability of the nerve cells overreaches their working capacity and a state of inhibition develops. When the stimulus is withdrawn the nerve centers return to their state of protracted, stationary excitation.

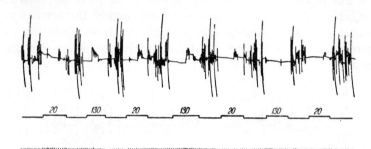

Fig. 31. Inverted stage. Inhibiting-balanced relationships.

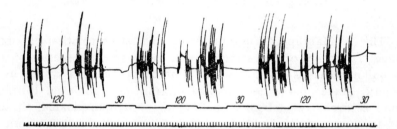

Fig. 32. Inverted stage. Inhibiting-paradoxical relationships.

An investigation carried out jointly with Fless showed that during the inverted stage distinctive relationships are observed between the intensity of the stimulus applied and the degree of its inhibiting action. Three forms of these relationships are observed:

a) During strong stimulation there is marked inhibition of activity, during weak stimulation there is very slight inhibition. We called such relationships "inhibiting-strength" forms (Fig. 30).

b) In response to weak and to strong stimulation the motor excitation is inhibited to an equal degree. These are "inhibiting-balanced" relationships (Fig. 31).

c) During strong stimulation inhibition is weak, during weak stimulation there is a strong inhibiting effect. We called these "inhibiting-paradoxical" relationships (Fig. 32).

Inversion of the strength relationships characteristic of parabiotic stages may thus be observed not only between the intensity of the applied stimulus and the response reaction of the nervous system, but also between the intensity of stimulation and its inhibiting action on the stationary excitation of the nerve centers.

The findings described above illustrate that the prolonged action of an external stimulus on the nervous system after the exhaustion of active inhibition leads to considerable excitation of the nerve centers—an excitation which may adopt an inert, protracted character. This excitation is restricted by limiting inhibition, which also leads to the appearance of reactions of different intensity in response to the applied stimuli, reactions which not only do not correspond to the law of strength relationships, but also in some cases have a profoundly inverted character.

When, in conjunction with Fless and Molodkina, in 1951 we began to study the pattern of development of the phases of limiting inhibition in the rats of our highly sensitive line, we rarely encountered individual animals with a well-marked inertia of the process of stimulation and with phases of limiting inhibition. Usually a decrease in the excitability of the animal's

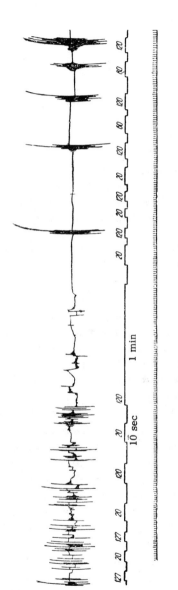

Fig. 33. Abolition of the state of excitation in a rat by limiting inhibition (caused by prolonging the action of the sound stimulus to 1 minute).

nervous system took place, despite the prolonged application of the sound stimulus.

Selection of the rats by the reaction of continued excitation, however, very rapidly (in the course of 1-2 years) led to the acquisition of this property by the individuals of our line (Krushinskii, 1959). At the present time the line of rats with which we are working is characterized both by high excitability and weakness of inhibition, and also by a considerable inertia of the process of excitation. During the prolonged, interrupted action of a sound stimulus the excitability of the nervous system of these rats is increased and maintained at a high level, and the motor excitation usually continues for a considerable time after the exposure to sound has ceased.

Having established the regular appearance of phases of limiting inhibition in rats when acted upon by a sound stimulus, we made an attempt to study the biological role of this inhibition in our model.

Pavlov showed that the inhibition of the higher nervous activity fulfills two functions: a coordinating function, which, together with the processes of excitation, ensures that the animal accommodates itself more fittingly to the external environment, and a protective function, designed to protect the nerve cells from overstrain and from functional disturbance under the influence of excessive stimuli.

Pavlov's concept of the protective role of inhibition has proved most valuable in the solution of theoretical and practical problems of the physiology and pathology of higher nervous activity. In the subject of protective inhibition, however, several points still require elucidation. For instance, the fundamental problem of the physiological mechanism of the therapeutic action of inhibition has not yet been explained. Is this therapeutic effect merely the result of the rest which the inhibition provides, or are the reparative processes which it stimulates of decisive importance?

As Fol'bort (1951) points out, the influence of inhibition as an agent promoting processes of recovery was, for Pavlov, only a general hypothesis, insufficiently supported by facts. In all the pathological cases which he analyzed, therefore, it has been overshadowed by the fundamental idea: the protective role of inhibition.

Experiments are described in the literature indicating the ability of inhibition to stimulate restorative processes. Fol'bort (1951), who used the concentration of the saliva as an index of the intensity of restorative processes, showed that after the application of an inhibiting stimulus, further stimulation caused the appearance of saliva of a higher concentration. Platonov (1952) showed that the strength of a muscle, exhausted on an ergograph to the point of inability to work, is restored two or three times as much in the course of hypnotic sleep for one minute as during rest for the same time in a waking state. The restoration of the respiratory ventilation after work also took place twice as quickly during sleep.

Our experimental model was found to be convenient for study of the problem of the physiological mechanism of the restorative role of limiting inhibition. With our ability to induce a state of deep limiting inhibition in an animal's nervous system, we studied the effect which it may exert on the subsequent functional state of the nervous system (Krushinskii and Fless, 1958). The inverted stage was particularly suitable for this purpose. By prolonging the action of the strong sound stimulus to 1-2 minutes instead of the normal 10 seconds, we were able at this stage to trace in a clear form the changes caused by the limiting inhibition in the state of marked excitation present in the animal.

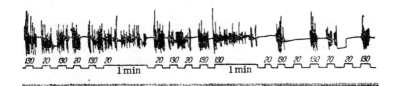

Fig. 34. Relationship between the restorative action of limiting inhibition and the strength of the stimulus.

It was found that after the parabiotic limiting inhibition had been strengthened and prolonged in this manner, there occurred a marked shift toward normalization of the functioning of the animal's nervous system: instead of a pathologically increased excitability with signs of explosiveness, protracted excitation, and inversion of the normal strength relationships, the animal became tranquil and the correct strength relationships between the magnitude of the stimulus and the response motor reaction were restored.

In Fig. 33 we show the reaction of a rat during the inverted stage. Prolongation of the sound stimulus to 1 minute at once restored the normal activity of the nervous system: the protracted excitation came to an end and the rat ceased to react to weak sound stimuli (20 and 60 v) and reacted only to stronger stimuli (80 and 120 v). Thus the induction of a profound limiting inhibition in the nervous system, developing as a result of prolongation of the sound stimulus, lowers its excitability and thereby normalizes its activity.

The same stimulus which causes the pathological state of the nervous system, by deepening the limiting inhibition normalizes its activity.

Further experiments showed that the efficacy of the restorative action of limiting inhibition is dependent upon its duration and depth. It will be clear from the kymograms shown in Fig. 34 that prolongation of the action of a weak sound stimulus (20 v) up to 1 minute had no appreciable effect on the functional state of the nervous system. Prolongation of the action of a strong stimulus (130 v), however, led to the restoration of the correct strength relationships. Consequently, the stronger the stimulus and the deeper the limiting inhibition caused by it, the more marked its restorative action.

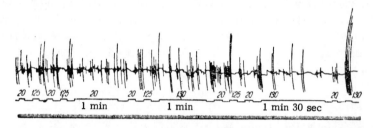

Fig. 35. Relationship between the restorative action of limiting inhibition and the duration of the stimulus.

In Fig. 35 we show the kymographic tracing of the motor reaction of a rat during the inverted stage. It will be seen from the kymogram that the prolongation of a weak (20 v) sound stimulus to 1 minute had no effect; prolongation of a strong sound stimulus (bell, 130 v), although it led to the emergence of the nervous system from the inverted stage, did not lead to the restoration of the correct strength relationships. However, when the same stimulus (bell, 130 v) was allowed to act for 1 minute 30 seconds, the correct strength relationships were restored. Consequently, the longer the nervous system is in a state of limiting inhibition, the more complete is the restoration of its normal activity.

A matter of decisive importance is the functional state of the nervous system when the action of the sound stimulus is prolonged. In Table 30 we compare the dependence of the effect of prolongation of the stimulus on the degree of abnormality of the activity of the nervous system.

It is clear from Table 30 that the prolongation of the sound stimulus lowers the excitability only if it is applied when the nervous system is in a state of pathological excitation, i.e., in the presence of balanced, paradoxical or inverted relationships, and in protracted

Table 30. Relationship between the Physiological Effect Caused by Prolongation of the Action of the Stimulus and the Functional State of the Nervous System

Functional state of the CNS before prolongation of the action of the sound stimulus	Number of cases	State of parabiosis after prolongation of the sound stimulus		
		Deepened	Unchanged	Weakened
Strength relationships	14	7	7	–
Balanced relationships.........	23	–	8	15
Paradoxical relationships.......	2	1	–	1
Inverted relationships	18	–	7	11
Protracted excitation..........	34	1	6	27

excitation. Prolongation of the sound stimulus when normal strength relationships are present, on the other hand, leads to a deterioration of the functional state of the nervous system, as is shown by the development of variations of phase or of protracted excitation.

Thus the same stimulus, acting on the nervous system in different functional states, may cause a diametrically opposite effect. When it deepens the existing limiting inhibition, the stimulus normalizes the functional state of the nervous system; when, on the other hand, limiting inhibition has not yet developed in the nervous system, it causes a state of pathological excitation (Fig. 36).

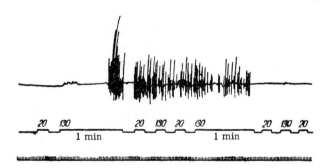

Fig. 36. Relationship between the effect of the action of a stimulus and the initial functional state of the nervous system.

This is an illustration of one of Vvedenskii's more important views (1886, 1901–1903), for according to him the functional state of the nervous system is not a predetermined property, but is the result of its preparation by the preceding action of a stimulus.

The lowering of excitability after the prolongation of the action of a sound stimulus is not the result of exhaustion of the process of stimulation. This may be seen from the fact that after the prolongation of the action of the sound stimulus, when the protracted excitation is replaced by normal strength relationships, the rat begins to react in response to the action of intensive sound stimuli with violent excitation (see Figs. 33, 35). It is clear that in this case the limit of the working capacity of the nerve cells has been exceeded, having been freed from the inhibition limiting it.

As an illustration of the active restorative role of limiting inhibition, experiments were conducted in which the effect of the prolonged action of a stimulus was compared with that of a corresponding period of rest (in the form of an interruption in the action of the stimulus).

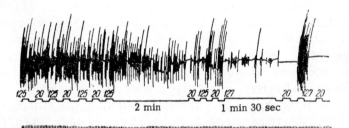

Fig. 37. The normalizing effect of prolongation of the stimulus, and its absence after an interruption in the action of the stimulus.

It is clearly shown in Fig. 37 that an interruption in the action of the stimulus for two minutes did not abolish the pathological excitation: the state of protracted excitation persisted in the rat. The action of the strong stimulus for a shorter time brought the rat out of this state, and led to establishment of the correct strength relationships between the intensity of stimulation and the magnitude of the motor reaction of the animal. This is evidence that the limiting inhibition, developing in this case, lowers the increased excitability and brings the nervous system out of its pathological state more quickly and fully than a brief rest.

This investigation thus showed that mechanisms of regulation against the action of excessively strong external stimuli exist in the nervous system.

One such mechanism is active inhibition (it reveals itself, as we pointed out above, at the very beginning of the action of the sound stimulus in the form of the inhibitory phase between the first and second waves of excitation), which we may regard as the "first line of defense." This inhibition depresses and restrains the excitation developing in response to the action of a strong external stimulus. It is unstable, however, and is readily exhausted. This inhibition can suppress excitation only for a relatively short period of time.

If a stimulus, having exhausted this inhibition, continues to act, the nervous system uses another defensive mechanism—its "second line of defense," i.e., limiting inhibition. This restricts the limit of excitability of the neurons, and thereby protects them from excessive excitation. Limiting inhibition, however, not only restricts the limit of excitation of the nervous system, but also actively removes the excitation. This latter happens when the action of the stimulus is sufficiently intensive and prolonged. As a result, the external agent which led to the development of the pathologically excited state of the nervous system normalizes its functional state.

This remarkable protective mechanism of the nervous system, created as the result of prolonged evolution, enables the nervous system to preserve its normal functional state, in spite of the mass of strongly acting stimuli.

The regulatory powers of this defensive system, however, are limited. When the inhibitory processes of the nervous system are no longer in a state to protect it from the action of excessive external stimuli, the body is threatened with serious consequences, which may even terminate in death. There is no doubt that the first place among the causes of death as a result of nervous trauma must be allotted to acute circulatory disturbances.

We have made a detailed study in our laboratory of the problem of acute circulatory disturbances as a result of functional trauma of the nervous system. These investigations have shown that in response to the prolonged action of a sound stimulus, marked excitation is observed in animals, which culminates in some cases in a severe shock-like state, and occasionally in the animal's death (Krushinskii, Pushkarskaya, and Molodkina, 1953).

As a standard we use an interrupted sound stimulus, acting for 15 minutes, in which weak and strong stimuli are applied for 10 seconds each, at intervals of 10 seconds. After this relatively long period of trauma, the action of the stimulus is interrupted for three minutes, after which the strong sound stimulus is reapplied for 1.5 minutes.

This system of application of sound stimuli has the result that during the interlude the excitability of the neurons of the brain is lowered, and the protective, limiting inhibition, is therefore removed, since this is evidently no longer physiologically necessary.

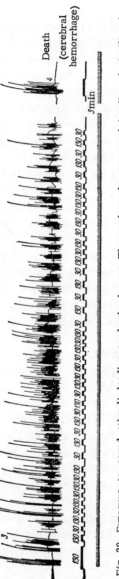

Fig. 38. Exposure to sound stimuli leading to death of a rat. The numbers denote convulsive fits: clonic (3) and tonic (4).

The subsequent application of the stimulus after this short rest, traumatizing the nervous system, leads to excitation which, not being restrained by the active process of inhibition (which was exhausted at the beginning of the "ringing"), and not being restricted by limiting, protective inhibition, follows a more intensive course and is more marked than before the interlude (Fig. 38). When the stimulus is applied in this way the animal's nervous system appears to be deprived of its protective, defensive inhibition. The marked excitation developing in this case in response to the action of the sound stimulus leads to an acute disturbance of the circulation, which may occasionally lead to the death of the animal.

Signs of pathological disturbances of the nervous activity (pareses, nystagmus, corneal opacity) appear relatively often, sometimes even before the interruption in the application of the sound stimulus, but they become particularly well-marked during the severe excitation after the three-minute interval.

Table 31 shows the incidence of pathological disturbances and of a fatal outcome in 165 rats to which sound stimuli were applied in accordance with the standard described above. Of the 21 animals dying as a result of the action of the sound stimulus, in 19 (90.5%) microscopic examination revealed hemorrhages in the brain.

Investigations carried out in our laboratory by Steshenko showed that a very brief (1.5 minutes) exposure to "ringing" could lead to a fall in blood pressure. For instance, in 22 animals, after motor excitation without a con-

Table 31. Incidence of Signs of Disturbances of the Functional State of the Nervous System and of a Fatal Outcome in Response to the Action of a Sound Stimulus

Type of disturbance	Absolute No.	%
Disturbance of muscle coordina-		
tion and pareses	56	33.9
Nystagmus	13	7.8
Lacrimation	7	4.3
Corneal opacity	19	11.5
Fatal outcome	21	12.7

vulsive fit, the blood pressure* fell on the average from 110 mm Hg to 94 mm Hg. In 74 rats in which the motor excitation culminated in a fit, the blood pressure fell still more: on the average from 107 to 83 mm Hg (difference 24 mm Hg). If no excitation reaction developed in the rats in response to the sound stimulus, the blood pressure was hardly affected. These investigations showed that a brief (1.5 minutes) exposure to a sound stimulus, causing excitation of the brain, leads to disturbance of the circulation.

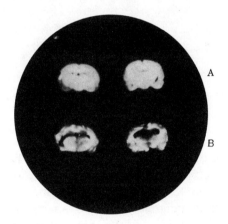

Fig. 39. Transverse section through the brain of a rat: A—the brain of a normal rat; B—of a rat dying from hemor-rhage into the brain.

After a longer exposure to sound (up to 15 minutes) the blood pressure showed no tendency to fall further. In spite of the fact

*The blood pressure was determined in the caudal vessels of the rats by the method of Sidney and Fridman (1941).

that no catastrophic fall of blood pressure was observed in the
rats, they nevertheless developed the typical picture of shock, with
marked depression of their reflex excitability, and a fall in the
body temperature, terminating sometimes in death, which usually
appeared either immediately after the action of the sound stimulus
or in the course of the next few days.

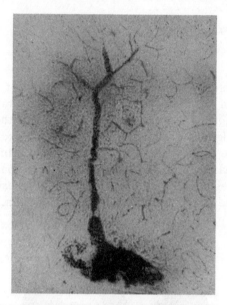

Fig. 40. Hemorrhages along a capillary in the
brain (region of the putamen of the lentiform
nucleus).

The degree of the fall of blood pressure can hardly be regarded
today as the foremost sign of the severity of a state of shock.
Serebryannikov (1951), for example, points out that the fall in
arterial pressure in peptone shock does not always correspond to
the degree of depression of the central nervous system. In his
opinion, the state of the reflex activity should be regarded as the
criterion of the severity of shock.

Histological investigations of the brain of rats dying after exposure to a sound stimulus, carried out by Shevchenko and, later,
by Oleneva (1955, 1958) and Svetukhina in Polyakov's laboratory,
showed marked disturbances in the vascular system of the brain.
In nearly all cases extensive hemorrhages were observed beneath
the pia mater or into the ventricles of the brain (Fig. 39), or hem-

orrhages along the individual capillaries of the brain (Fig. 40). The hemorrhages occurred as a result of diapedesis through the walls of the dilated cerebral capillaries, which were in a state of paresis (Fig. 41).

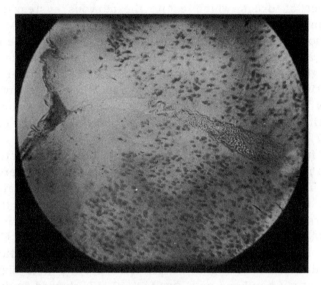

Fig. 41. Severe dilatation of a capillary in the cerebral cortex.

In our experiments the hemorrhagic component of the state of shock thus stands out clearly in neurogenic shock. The hemorrhages taking place into the pia mater, the ventricles of the brain and the brain substance, and also found in other organs (lungs), are evidently the main cause of death of an animal during the development of neurogenic shock.

The changes occurring in shock found by Koreisha clinically (1955) and in physiological experiments (1957) are extraordinarily reminiscent of the pathological picture which develops in our experimental rats in response to severe excitation of the brain, caused by the action of an intensive sound stimulus.

Experiments conducted in our laboratory showed that a state of shock, terminating fatally as a result of acute circulatory disturbance, may be produced not only by direct trauma to the nervous system, but also as a result of the severe excitation of the brain caused by the action of a strong exteroceptive stimulus.

Having established the pattern and conditions of development of states of shock and hemorrhage developing after nervous trauma,

156   Chapter 4

we set out to discover the physiological conditions of the body which
encourage or prevent this pathological process.

Since the state of shock and hemorrhage in the rats develops
as the consequence of severe excitation of the brain, we studied
the effect of drugs which excite the nervous system (caffeine).
These investigations (Krushinskii, Pushkarskaya and Molodkina,
1953) showed that the preliminary administration of caffeine to the
rats (in a dose of 10-20 mg/100 g body weight) considerably increases the severity of the states of shock and the number of cases
terminating fatally. For instance, of 49 animals receiving preliminary caffeine, 16 (32.6%) died after nervous trauma by the
sound stimulus (in a control group not receiving caffeine the figure
was 12.7%).

The functional state of the glands of internal secretion has a
significant influence on the incidence of fatal outcomes after nervous
trauma. The role of the parathyroid glands has been explained. The
study of the parathyroid glands was dictated by the fact that these
glands regulate the level of the concentration of calcium in the
blood. Many investigations, starting with the work of Cohnheim
and Arnold, have demonstrated the fact that calcium deficiency in
the medium surrounding the capillaries may lead to diapedesis.
Subsequently Chambers and Zweifach (1947) showed by direct experiments that the permeability of the walls of a vessel is dependent
on the calcium content of the intercellular cement.

A low calcium content of the diet (0.004-0.02%) leads to the
development of multiple hemorrhages in several organs of animals
(Greenberg et al., 1939; Boelter et al., 1941). Hemorrhages are
frequently observed in the central nervous system. They have been
detected in the cerebellum, the medulla, the brainstem, and the
lateral ventricles. These authors conclude from their investigations
that a low calcium content in the diet predisposes animals to hemorrhages, which may be the result of a disturbance of the permeability of the vessel walls.

It is also currently accepted in the medical literature that a
deficiency of calcium in the body may be a significant cause of
hemorrhage. Bogolepov (1950), when discussing the cause of the
high incidence of cerebral hemorrhage in Leningrad during the
blockade, stated: "It is possible that the development of the hemorrhagic syndrome in wartime was favored by the hypocalcemia,
hypoproteinemia, and deficiency of vitamins C and, in particular,
K, present in many hypertensive patients."

Investigations carried out on dogs by Polushkina (1951) showed that after parathyroidectomy traumatic shock develops much more quickly. She observed a lowering of the excitability of the vagus nerves, an increase in the sensitivity of nerves to mechanical stimulation, an increase in the blood pressure in response to stimulation of nerves, and other physiological changes.

Table 32. Number of Deaths in Parathyroidectomized and
Control Rats after Trauma with a Sound Stimulus

Results of trauma with sound stimulus	Parathyroid-ectomized animals	Control animals
Number of rats dying.............	22	11
Hemorrhages found in the brain when examined macroscopically.........	18	7
Number of rats dying, as % of animals tested.....................	59.5	28.9

Our experiments to investigate the role of the parathyroid glands in the development of shock and hemorrhagic disorders were carried out on 37 parathyroidectomized and 38 control animals (Krushinskii, Pushkarskaya, and Molodkina, 1953). Animals after extirpation of the parathyroid glands (and the parallel series of control animals) were exposed to the action of a sound stimulus three times (at periods of one to three weeks after the operation) (Table 32).

It may be seen from Table 32 that nearly twice as many parathyroidectomized as control animals died as a result of nervous trauma. It is not only an increase in the number of deaths among parathyroidectomized animals, however, which characterizes the severity of the state of shock which develops. We also observed a considerable increase in the number of cases of disturbance of coordination of movement, of paresis and of nystagmus in the experimental group, compared with the controls.

The reports in the literature and our own experimental investigations thus demonstrate the role of the blood calcium level in the development of shock and hemorrhagic states. The absence of the parathyroid glands predisposes to the development of a more severe state of shock.

In the previous chapter we showed the sex determined differences in the fundamental properties of the nervous activity and in the unitary reactions of behavior of animals. The nervous system in the male is more resistant to the action of strong stimuli than in the female. In shock and hemorrhagic states the opposite results were obtained.

The course of shock in the male is much more severe and it terminates more often in death than in the female. For instance, of 102 males, 16 (15.7%) died as a result of trauma with a sound stimulus, and of 63 females only five (7.9%) died.

Later experiments showed that castration of males increases their resistance to the development of shock or hemorrhage after exposure to a sound stimulus.

The hormone of the thyroid gland has a particularly great influence on the development of shock or hemorrhagic states as a result of nervous trauma. The role of the thyroid hormone in the development of various forms of shock has been the subject of considerable research. In 1938 Burdenko (in a joint paper with Smirnova) pointed out that since the mediators of the nervous system play an essential role in the development of shock, an artificial increase in oxidative processes by means of thyroidin must predispose to the development of shock.

Subsequent experimental research showed that a hyperthyroid state increases the predisposition to various forms of shock. Petrov (1939, 1947) preliminarily injected rabbits with thyroidin and found that they became more sensitive to electric shock than normal animals.

Obvious changes in the sensitivity of dogs to peptone shock in hyperthyroidism were obtained by Baturenko (1940, 1941). He found that the preliminary administration of thyroidin hastens the onset and increases the severity of peptone shock. The investigations of Kovshikov (1950) showed that the preliminary administration of thyroidin to rabbits increases the severity of burn shock. Hyperthyroidization increases the sensitivity of cats to traumatic shock (Belichenko, 1952, 1953).

Kudienko (1957) obtained different results in anaphylactic shock. According to this worker, thyroidectomy aggravates the symptoms of anaphylactic shock.

Investigations conducted in our laboratory showed that the preliminary administration of thyroidin to rats (from 0.025 to 0.1 g) during the 10–14 days before the experiment increases by many times the mortality arising as a result of nervous trauma (Krushinskii and Dobrokhotova, 1957) (Table 33).

Besides the sharp increase in the number of fatal cases, a noteworthy feature was the far more severe state of shock in all

Table 33. The Influence of Hyperthyroidization on
the Mortality Rate from Nervous Trauma

Group of animals	Number of rats in experiment	Number which died	
		Absolute number	%
Receiving thyroidin	103	69	66.6
Controls	108	9	8.3

the animals receiving thyroidin, compared to the controls, both
during the action of the sound and subsequently. In the majority of
the rats receiving preliminary thyroidin the signs of severe shock
arose from the very first minutes of the action of the sound stimulus,
and death ensued before the exposure to sound had ceased.

Full agreement with the results of these experiments is shown
by data obtained in animals after thyroidectomy (in control animals,
only the parathyroid glands were removed*).

As a result of trauma from a sound stimulus many more thyroidectomized than control rats died. The difference between the mortality rates of the two groups came close to the requirements of
statistical significance: it was 2.21 times the probable error.

Blocking the function of the thyroid gland with methylthiouracil
gave similar results (Dobrokhotova, 1957). The preliminary administration of methylthiouracil (from 12 to 60 mg daily for 10−14
days) lowers the mortality rate from nervous trauma in experimental
animals in comparison with controls (among control animals the
mortality was 2.7 times that among animals receiving methylthiouracil).

During exposure to sound the state of the animals of the control
group was more serious than that of the experimental group. For
instance, among the rats preliminarily receiving methylthiouracil,
motor disturbances (pareses, disturbances of the coordination of
movements) were observed in 31.3% of the animals, and among the
control group, in 54.3%. (Statistical treatment of the incidence of
these disturbances showed that the difference between the two
groups was 4.8 times the probable error, i.e., that it was definitely
significant.)

Experiments using both hyper- and hypothyroidism thus demonstrated the very considerable influence of the thyroid hormone
content on the outcome of neurogenic shock. An increased content

*It is practically impossible in rats to remove the thyroid gland without the parathyroids.

of the hormone greatly increases the number of fatal results in nervous trauma.

A detailed analysis of the state of excitability and the inhibitory function of the nervous system in rats traumatized by a sound stimulus was made by Dobrokhotova (1958). Her investigations showed that the state of excitability of the nervous system (as assessed by the reaction of the rat to the sound stimulus) of hyperthyroidized rats at the beginning of the nervous trauma is absolutely indistinguishable from the state of nervous excitability of control animals. In hyperthyroidized animals, during the period of nervous trauma when associated with a state of excitation of the brain, however, a tendency is observed toward a lowering of the excitability of their nervous system as compared to the controls. This indicates that the increased rate of mortality resulting from sound trauma in hyperthyroidized animals is not determined by the increased excitability of those structures of the brain which are concerned in the pathological process.

According to findings obtained in our laboratory, the cause of the difference in mortality between normal and hyperthyroidized animals must be sought in acute disturbances of the circulation. Data in the literature (Petrov, 1939; Kovshikov, 1950, and others) and the results obtained in our laboratory show that hyperthyroidism leads to a rise in the arterial blood pressure. As shown by Steshenko's investigations, in rats receiving thyroidin (0.025–0.05 g), the average blood pressure was 154.4 ± 4.68 mm Hg (21 animals), and in the control group of animals (20 animals) it was 119.0 ± 2.8 mm. The difference of 35.4 ± 5.49 mm Hg, being 6.45 times the probable error, confirms that the conclusion was correct.

At the same time Steshenko's investigations showed that the mortality rate among hyperthyroidized rats having a high initial level of the blood pressure was greater than that among those having a lower blood pressure (Table 34). The increase in the blood pressure as a result of hyperthyroidization is evidently one of the causes of the higher mortality among these animals as a result of neurogenic shock.

Besides increasing the blood pressure, hyperthyroidization leads to an increase in the permeability of the vessels (Mogil'nitskii, 1949; Kerekesh, 1955; Bolotova, 1956; and others). The investigations of Steshenko, performed in our laboratory, showed that in hyperthyroidized rats the permeability of the capillaries of the skin was increased in comparison with control animals.

Table 34. Initial Value of the Blood Pressure in Hyperthyroidized
Rats Surviving and Dying After the Action of a Sound Stimulus

Group	Number of animals	Mean value of blood pressure, in mm Hg	Difference	Significance of difference
Survived.......	79	121.11	16.36 2.24	3.88
Died	55	137.47		

An increase in the arterial pressure and an increase in the
vascular permeability are significant to the development of neurogenic shock.

As a result of nervous trauma a considerable fall of the blood
pressure takes place. Steshenko showed that in a group of hyperthyroidized rats (18 animals) after only 1.5 minutes of exposure
to the sound stimulus the arterial pressure fell by 37 mm (from
128 to 91 mm). During continued nervous trauma, no further drop
in the arterial pressure was observed. In animals which died a
short time after trauma from the sound stimulus, however, this
fall was greater than in those which survived.

It must be pointed out that in the groups of rats which died there
was a greater fall in the body temperature (by 0.9°C) than in those
which survived (by 0.2°C).

A greater fall in the blood pressure of hyperthyroidized animals,
as found in neurogenic shock, has also been observed in other forms
of shock by other authors (Rosin, 1938; Petrov, 1939; Kovshikov,
1950; Baturenko, 1951; Belichenko, 1953).

Bogoyavlenskaya (1957, 1958) showed that after excitation and
epileptiform fits, caused by sound stimulation for 1.5 minutes, the
blood prothrombin of animals decreases on the average by 22%
(during the 15 minutes after cessation of the action of the sound
stimulus). Under these circumstances the relative concentration
of thrombotropin and heparin in the blood remained unchanged.
After more prolonged trauma to the nervous system (lasting 15
minutes), leading in some cases to a state of shock and death from
cerebral hemorrhage, an average decrease of 26% in the blood
prothrombin was observed. Exposure of rats not in a state of excitation to the action of sound also caused a fall in the blood prothrombin level.

Bogoyavlenskaya concludes that the decrease in the blood prothrombin concentration in rats sensitive to the sound stimulus is
not large enough to be a decisive factor in the development of
cerebral hemorrhage.

It is obvious that the disturbance of the coagulation of the blood
is not the chief mechanism of production of shock and hemorrhagic
conditions which develop in rats after nervous trauma, but is merely
one of the conditions favoring the development of shock.

Table 35. Result of Experiments in which Carbon Dioxide was
Inhaled Throughout the Period of Nervous Trauma to Rats Preliminarily Sensitized with Caffeine

Conditions of experiment	Number of animals	Mean blood pressure (in mm Hg)		Number of disturbances of movement during experiment		Number of deaths during or after experiment	
		Before experiment	After experiment	Absolute	%	Absolute	%
Inhalation of CO_2 (7%)	20	119	100	11	55.0	1	5.0
Normal atmosphere (control)	21	119	85	20	95.2	7	33.3

Investigations carried out in our laboratory (Krushinskii, Steshenko, and Molodkina) showed that the inhalation of carbon dioxide
by an animal during nervous trauma leads to a decrease in the
intensity of the manifestations of developing shock and to a reduction in the mortality rate. In these experiments the animals were
placed in a chamber containing from 7 to 13% of carbon dioxide*
either throughout the period of action of the sound stimulus, or at
certain moments of its application. The number of animals used
in the experiments was 304—169 experimental and 135 control. As
a preliminary step, a proportion of the animals was sensitized to
the development of shock by means of the injection of thyroidin
or caffeine.

As an example of the influence of inhalation of carbon dioxide
throughout the period of trauma by the sound stimulus, we may
cite the results of experiments with rats preliminarily sensitized
with caffeine (100 mg/kg caffeine sodium benzoate was administered
to both experimental and control rats 30–40 minutes before the
beginning of action of the sound stimulus) (Table 35).

*The concentration of carbon dioxide was determined by Haldane's method.

Table 35 shows differences both in the degree of fall of blood pressure and in the mortality rate between animals kept in a normal atmosphere and those kept in an atmosphere with an increased concentration of carbon dioxide. The results of the experiments in all the series showed that the mortality rate among the rats spending the entire period of trauma by the sound stimulus in an atmosphere containing an increased concentration of carbon dioxide was 11.8%; the mortality rate among the animals kept in a normal atmosphere was 37.9%. The optimal carbon dioxide concentration for the prevention of shock and hemorrhagic states was 7%.

In spite of the fact that the inhalation of carbon dioxide lowers the mortality from neurogenic shock, it is at present difficult to define accurately the physiological mechanism by which this effect is produced. The work of Kosman and Damour (1956) and of Fless and Steshenko in our laboratory showed that inhalation of carbon dioxide leads to the weakening, and in high concentrations to the total abolition of the reaction of excitation and fits in rats in response to sound stimulation. It is possible that the reduction in the intensity of the shock states is due to a lowering, under the influence of CO_2, of the excitability of those divisions of the central nervous system which are responsible for the development of the pathological excitation. It may be admitted, however, that since carbon dioxide strongly dilates the vessels of the brain, it prevents the development of the primary spasm of the cerebral vessels which, as shown by Bella-Bella (1954), occurs in rats subjected to electric shock. The inhalation of carbon dioxide, by diminishing the subsequent paresis of the capillaries, thus prevents hemorrhages and lessens the severity of the state of shock.

So far we have examined the pathological states developing in rats as a result of the

Fig. 42. Motion picture recording of a myoclonic spasm.

single application of a sound stimulus. These pathological changes, which develop acutely, disappear after a short time and are not reflected in the subsequent reactions of the animal in response to the sound stimulus. The systematic application of the sound stimulus may, however, lead to chronic disorders of nervous activity.

We have observed the development of hyperkineses in the form of tic-like myoclonic spasms, which constantly occur during the action of a sound stimulus, and in conjunction with Molodkina we have studied their physiological nature. The myoclonic spasms begin, as a rule, in the muscles of the head; they are particularly noticeable in the eyelids and auricles; the convulsions then pass to the forepaws: the rat sits on its hindpaws and makes interrupted twitching movements of the head and forepaws (Fig. 42). When these myoclonic spasms are more marked, they may also extend to the hindlimbs. In this case the animal may even fall on its side.

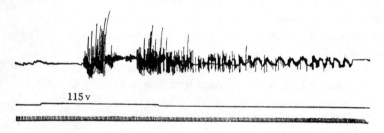

115 v

Fig. 43. Kymographic tracing of myoclonic spasms appearing during the inhibitory period of the reaction.

These myoclonic spasms are never observed in rats exposed to the action of sound stimuli for the first time; once they have appeared, however, they increase in intensity from one experiment to the next, and develop into a permanent pathological phenomenon. Myoclonic spasms never appear during the latent period, nor at the beginning or in the middle of the first wave of excitation. They appear as a rule in the inhibitory phase of the reaction (Fig. 43), and sometimes at the end of the first wave of excitation. Of 42 rats, for instance, the myoclonic spasms first developed during the period of inhibition, and in only two rats did they begin at the end of the first wave of excitation, immediately before the beginning of the development of the period of inhibition. When they began in the period of inhibition, they sometimes spread to the beginning of the second wave of motor excitation.

Observations on the pattern of appearance of the myoclonic spasms have shown that they are in some way connected with inhibition. In our investigations, carried out jointly with Levitina and Molodkina (Krushinskii, et al., 1959) it was discovered that the time required for the restoration of inhibition, when interrupted by the onset of excitation, becomes longer in most rats from one experiment to the next. This indicates that inhibition is weakened as a result of chronic trauma to the nervous system by the sound stimulus. Myoclonic spasms appeared against the background of the functional weakening of the process of inhibition.

From the facts described above we formulated the following working hypothesis concerning the nature of this pathological phenomenon. Myoclonic spasms develop as a result of the chronic exhaustion of inhibition, in consequence of the systematic overstraining of this process during the action of the sound stimulus. The weakened process of inhibition is incapable of resisting the impulses of excitation coming from pathologically excited centers of the brain.

A decisive fact in support of this hypothesis is that myoclonic spasms appear more often during the shorter intervals between two ringings (on one day of the experiment), when the inhibition cannot be adequately restored, than during the longer intervals between the action of the sound, during which a more complete restoration of the inhibition takes place.

The experimental method was as follows. On the day of the experiment, rats in which myoclonic spasms appeared were exposed twice to the action of the sound stimulus. The first time, the bell was switched off as soon as a period of inhibition developed in the animal. After a short interval (for example, 10–15 minutes) the rat was again subjected to the action of the sound stimulus for 1.5 minutes. If during this second exposure the animal developed myoclonic spasms, on the next day of the experiment (the experiments were carried out on alternate days) the interval between the two ringings was lengthened; if the rat then gave a reaction without myoclonic spasms, the interval between the first and second tests was shortened.

If, in fact, myoclonic spasms appeared when the process of inhibition was exhausted, then in the case of the shorter intervals during which there was less chance of recovery of the inhibition than during the longer intervals, myoclonic spasms must develop more often than during the longer intervals. Experiments were carried out in which shorter and longer intervals alternated (Table 36).

Table 36. Relationship between Myoclonic Spasms and the Interval
Between Ringings

Number of experimental rats	Number of comparable pairs of intervals	Appearance of myoclonic spasms in comparable pairs, depending on length of interval	Number of comparable pairs of intervals	
			Absolute number	%
51	216	Myoclonia during shorter intervals but not during longer	162	75
		Myoclonia during longer intervals but not during shorter	54	25

The results given in Table 36 clearly demonstrate that myoclonic spasms develop much more often during the shorter intervals
between ringings than during the longer intervals. This suggests
that weakness of the process of inhibition (being unable to recover
during the short time after the previous ringing) is a condition
favoring the development of myoclonic spasms. In the course of the
longer interval, the process of inhibition is able to recover to such
an extent that the myoclonic spasms develop less frequently.

It is interesting to note that the difference between the intervals
during which myoclonic spasms develop and those in which they are
absent may be very small. In rat No. 2325, for example, this difference was 2 minutes (Table 37).

Table 37. Appearance of Myoclonic Spasms Depending on the Intervals between the First and Second Ringing

Date	Length of interval between first and second exposures of rat to sound in the experiment (in minutes)	Result of second application of sound stimulus (after interval)
Sept. 20	35	Reaction with myoclonic spasm
Sept. 22	50	Two waves of excitation without myoclonic spasm
Sept. 27	35	Reaction with myoclonic spasm
Sept. 29	50	Two waves of excitation without myoclonic spasm
Oct. 2	37	Two waves of excitation without myoclonic spasm
Oct. 4	37	Two waves of excitation without myoclonic spasm
Oct. 6	35	Reaction with myoclonic spasm

It is clear from Table 37 that when the intervals between the periods of sound stimulation were 35 minutes, myoclonic spasms developed during the second application. When the intervals between the two periods of stimulation were 37 minutes or longer, no myoclonic spasms were observed. In another rat the corresponding intervals were 58 and 60 minutes.

If the sound stimuli were systematically applied it was necessary to increase the time between the two applications on a single day of the experiment in order to obtain reactions without myoclonic spasms. This demonstrates that there is a gradually progressive weakening of the process of inhibition, on the state of which depends the appearance of this chronic pathological phenomenon.

Table 38. Change in the Incidence of Myoclonia under the Influence of Sodium Bromide

Action of sound stimulus tested	Number of tests	Reaction with myoclonia		Reaction without myoclonia	
		Absolute number	%	Absolute number	%
Before injection of bromide.	70	58	82.9	12	17.1
On the day of injection of the bromide and on the next day	104	45	43.3	59	56.7
On the 2nd-4th day after injection of bromide	38	21	63.6	12	36.4
On the 5th-7th day after injection of bromide	22	18	81.8	4	18.2

The relationship between the development of myoclonic spasms and the functional state of the process of inhibition indicated that it might be possible to prevent this phenomenon or to diminish its frequency by means of a bromide.

An experiment was carried out on eight rats with chronic myoclonic spasms. Sodium bromide was injected 30–40 minutes before the beginning of the trail, in doses of 15–45 mg (the doses were selected individually for each rat). Bromide was given for periods varying from a few days to 1–1.5 months (Table 38).

Table 38 shows the considerable decrease (from 82.9 to 43.3%) in the incidence of reactions accompanied by myoclonia on the day of injection of bromide and on the following day. By the 2nd–4th day after administration of bromide has ceased, however, the incidence of reactions with myoclonia has increased to 63.6%, and

by the 5th–7th day it has practically regained its original value
(81.8%) (Molodkina, 1956).

The influence of bromide on the incidence of myoclonia clearly
demonstrates the importance of the state of inhibition for the development of this pathological phenomenon. These and other investigations into the physiological analysis of the nature of myoclonic
spasms give evidence of the leading role of a weakened process of
inhibition in the production of this condition. The exhaustion of
inhibition as a result of its systematic overstrain is the foremost
cause of the development of myoclonic spasms.

Our investigations thus show that myoclonic spasms are based
upon functional changes analogous to those observed in the development of the neuroses (a weakening of the process of inhibition).

Obsessive states during various forms of neurosis in dogs were
studied in Pavlov's school by many of his collaborators and followers
(Podkopaev, 1926; Rikman, 1932; Kleshchov, 1938a; Petrova, 1939;
Kupalov, 1949; Usievich, 1949; Pavlova, 1949; Yakovleva, 1949;
Dolin and Zborovskaya, 1952; Vetyukov, 1936; Glisson and Shumilina,
1941; Alekseeva, 1949; Apter, 1952).

Pavlov regarded obsessive, stereotyped movements as the
manifestation of the "stasis" or "inertia" of the process of stimulation (Pavlov's Wednesdays, 1949, Vol. 1). He considered that the
cause of this stasis was overstrain of the process of stimulation.

He considered, however, that the primary cause of the pathological inertia of the process of stimulation was a weakening of
the process of inhibition. He stated: "It is natural to dwell on the
hypothesis that pathological inertia of the process of excitation is
a secondary phenomenon, resulting from the weakening of the
process of inhibition."* Our analysis of the physiological mechanisms of myoclonic spasms fully confirms these statements of Pavlov.

In our laboratory we have also studied some other sequelae of
acute excitation of the brain caused by a sound stimulus. Investigations by E. V. Gura,† under the direction of Professor M. Ya.
Fratkin, showed that after exposure of only 1.5 minutes to the
action of sound a slight decrease of intraocular tension takes place,
and if the duration of the exposure is increased, the decrease of
intraocular tension becomes very marked. In animals subjected to
the action of an interrupted sound stimulus for eight minutes, the

---

*I. P. Pavlov. Pavlov's Wednesdays [In Russian] Academy of Sciences, USSR Press, v. 2,
p. 12.
†From the Helmholtz Institute of Eye Diseases.

intraocular tension reaches 14–16 mm Hg, and after the action of such a stimulus for 15 minutes it falls to the characteristic tension of the "cadaver" eye. The cause of this phenomenon is the inhibition which develops immediately after the excitation from the action of the sound stimulus.

A similar change in the intraocular tension is observed in animals preliminarily rendered immobile with curare. This demonstrates that it is not the result of the animal's prolonged activity, but is associated with excitation of the central nervous system. The intraocular tension returns to its original level after 20–180 minutes, depending on the degree of trauma to the nervous system.

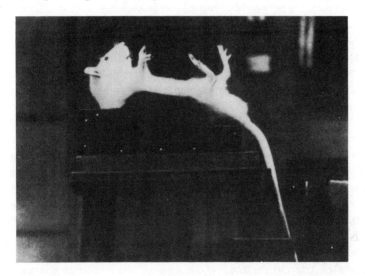

Fig. 44. Cataleptoid state in a rat after the motor reaction to a sound stimulus.

As a result of these investigations it is possible to study the connection between the intraocular tension and the functional state of the nervous system. This is essential for the elucidation of certain aspects of the development of that serious disease, glaucoma.

As we have previously pointed out, one of the consequences of the acute excitation or convulsive fit caused in rats by a sound stimulus is a cataleptoid state. In the period following the reaction of excitation or the epileptiform fit a profound stupor is observed, against the background of which a cataleptoid state may develop from the very outset. During this period the rats may adopt most extraordinary postures (Fig. 44).

Research conducted in our laboratory (Prokopets, 1958, 1958a) showed that a major factor determining the appearance of a cataleptoid state is the preceding excitation caused by application of the sound stimulus. In rats exposed to the action of a sound stimulus but not displaying motor activity, stupor and catalepsy are never observed. In the period of development of deep cataleptic stupor, disturbances of sensation are seen, in addition to the characteristic changes in the muscle tone.

Because of the general akinesia and areflexia manifested in the initial stage of catalepsy it may be postulated that this state is based upon a wide radiation of inhibition, affecting not only the motor area of the cortex, but also the subcortical centers. To the picture of acute changes in the somatic sphere are added disturbances of vegetative functions: the profuse secretion of nasal mucus, lacrimation, urination, changes in respiration, and so on.

It must be pointed out that other external stimuli cannot bring the animal out of this state of profound stupor and catalepsy. Only the sound stimulus can do this. It would thus seem that the stimulus which induced the cataleptoid state is also the specific stimulus extricating the animal from it.

Parallel with the increase in the intensity of the reaction (from motor excitation alone to excitation with a clonic convulsion), the duration of the cataleptoid state* increased on the average from 11.89 ± 0.97 to 18.17 ± 1.37 minutes. This increase is statistically significant. It is interesting to observe that after the strongest convulsion (assessed as 4 points), during which the animal is in a state of convulsions with tonic contraction of all its muscles, the duration of the cataleptoid state was shorter than in the remaining cases (8.58 ± 0.91 minutes).

The investigations showed that the duration of the cataleptic state is also dependent on the character of the waves of motor excitation. When the excitation reaction is of the single-wave character, the cataleptic state is much more prolonged than when it is formed of two waves, and this difference is statistically significant.

Further experiments showed that if there is pathological inertia of the process of stimulation, the duration of the state of catalepsy in the animals is considerably prolonged.

*The duration of the cataleptoid state was determined by the full recovery from the "lying-on-its-back" position to the postural reflex of standing up.

In complete agreement with the foregoing findings are the results of experiments in which the excitability of the nervous system was increased by means of the administration of caffeine, which increases the duration of the protracted excitation. It was found that after the administration of caffeine the inertia of the process of stimulation was increased, on the average, more than fivefold, and the duration of the cataleptoid state rose from 15.25 ± 2.50 to 25.69 ± 2.36 minutes. The difference of 10.44 ± 3.44 minutes is statistically significant; it is more than three times the probable error.

The systematic (daily) traumatization of the animal's nervous system by means of a sound stimulus, which, as we showed above, weakens the process of inhibition, leads to a progressive increase in the duration of the cataleptoid state. The mean duration increased from 12 minutes (second trial) to 40.3 minutes (sixth trial).

Pavlov regarded catatonic inhibition as the manifestation of a characteristic form of self-defense.

Prokopets (1958a) investigated the problem of the protective and restorative role of cataleptoid inhibition. He showed that the secondary application of a sound stimulus 2–3 minutes after the end of the epileptiform reaction, i.e., while cataleptic inhibition was developing, leads to a decrease in the excitability and to a weakening of the motor activity of the animals. This is shown by the considerable lengthening of the latent period (on the average by 18.15 ± 4.43 seconds) and by the considerable weakening of the intensity of the epileptiform reaction (from 2.71 ± 0.22 to 1.11 ± 0.13 points). These differences were statistically significant. It should be added that in 18.5% of animals the motor reaction in general was absent.

These findings are an illustration of the fact that the cataleptoid inhibition, developing after acute excitation (and the convulsive fit), plays a protective and restorative role.

The problem of the morphological structures of the brain involved in the process of excitation during the action of the sound stimulus was studied in two directions at our laboratory: Kotlyar (1958, 1959) extirpated different divisions of the brain (usually the cerebral hemispheres) and studied the influence of decortication on the subsequent pattern of development of a pathological reaction in the rats; Vasil'eva, Semiokhina, and Gusel'nikova studied the action potentials in different divisions of the brain in rats during the excitation reaction, the convulsive fits, and the myoclonic spasms.

There is no convincing information in the literature regarding the localization of the nervous processes developing in rats during exposure to sound. The results obtained by extirpation of different divisions of the brain are contradictory.

Weiner and Morgan (1945), for instance, performed bilateral extirpation of the motor, frontal, and auditory regions of the cortex, and observed a considerable decrease in the number of fits; whereas Beach and Weaver (1943) observed an increase in the sensitivity to sound stimulation after the unilateral extirpation of more than 90% of the neopallium, large areas of the hypothalamus, and the corpus striatum.

Kotlyar showed in more than 100 rats that partial decortication does not alter the character of the reaction of excitation and epileptiform fit. Similar results were obtained by Van Binh (1958) after removal of the anterior part of the cortex.

After the extirpation of more than 90% of the cortex the principal indices of the reaction (magnitude of latent period, character of waves of motor excitation, and intensity of convulsive fit) were unchanged; only in rats with a double wave of excitation was a slight increase in the reaction observed.

These experiments show that the cerebral cortex can hardly be responsible for the development of the excitation reaction and the convulsive fits in rats in response to the action of the sound stimulus. The chronic pathological condition (hyperkineses, appearing in the form of myoclonic spasms) may be completely abolished by extirpation of the cortical part of the cerebral hemispheres. The investigations of Kotlyar showed that in rats, after decortication, the chronic myoclonic spasms were completely lost (Fig. 45).

To investigate the region of cerebral localization of the myoclonia, bilateral extirpation of the motor, cutaneous-kinesthetic, auditory, and optic analyzers was carried out (according to V. M. Svetukhina's cytoarchitectonic chart of the brain). After extirpation of the greater part of the grey matter within the limits of the motor or cutaneous-kinesthetic analyzers, in which are concentrated the largest number of large pyramidal cells in the lower level of the cortex in rats, the myoclonic spasms became less severe, affecting only the muscles of the head and the forelimbs. Bilateral extirpation of the auditory and optic analyzers did not alter the character of the course of this hyperkinesis.

In rats in which myoclonic spasm had completely disappeared after extirpation of the sensomotor area of the cortex, although these spasms could be produced again, they followed an abortive

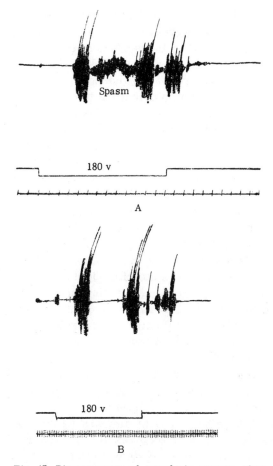

Fig. 45. Disappearance of myoclonic spasms after removal of the cerebral cortex. A) Reaction of an intact rat;   B) reaction of the same rat after decortication.

course: the animal merely shook itself and occasionally gave convulsive spasms of the limbs.

Investigations in which the cerebral cortex was extirpated thus showed that two pathological reactions could be distinguished: motor excitation and convulsive fits (which occur mainly in the subcortical structures of the brain), and myoclonic spasms (which mainly involve the participation of the neurons of the cortex).

The investigations of the electrical activity of the brain are in full agreement with the results of extirpation of the cortex. They assisted in the study of those concrete structures of the brain in which pathological changes take place during the action of a sound stimulus. Electrophysiological investigations enable us to judge not only the localization of the focus of excitation, but also the extent of its spread along the various structures of the brain.

Bureš (1953, 1953a) recorded the electrical potentials of the cerebral cortex during pathological excitation arising in response to the action of a sound stimulus. In order to increase the sensitivity of the rats, which were held stationary on the bench, he administered cardiazol. During the clonic stage of the fit he recorded the development of a typical epileptoid activity in the form of slow waves and "peak-wave" complexes with a frequency of three per second. He concluded from the changes in the pattern of the bioelectrical activity of the brain during the convulsive fit that at this period the cortex is in a state of inhibition. Bureš's investigations took the form of acute experiments.

In our laboratory Vasil'eva (1957) developed a method of restraint of rats which permits the electrical activity of the brain to be recorded in chronic experiments during the development of motor excitation and convulsive fits in rats in response to a sound stimulus. In order to sensitize the animals, parathyroidectomy was carried out which, by weakening the process of inhibition, facilitated the manifestation of the fit.

Vasil'eva found marked changes in the action potentials in the motor area of the cortex during myoclonic spasms. These investigations were subsequently continued by Semiokhina (1958) and Gusel'nikova (1958, 1959) also in animals with long-standing implantation of electrodes in the cortex and in various regions of the subcortex (the striopallidary system, the internal geniculate body, the rhinencephalon, the hypothalamic region, the medulla oblongata, and the cerebellum).

Gusel'nikova found that the greatest changes in the electrical activity during a convulsive fit took place in the medulla oblongata (the region of the vestibular and auditory nuclei and the reticular formation at the same level) (Fig. 46A). These changes took the form of epileptoid discharges of high amplitude, and moreover, even during the latent period it was sometimes possible to observe a marked increase in the frequency of the original background waves. During the tonic stage a "peak-wave" complex was recorded, which did, in fact, appear in this stage of the fit. A "peak-wave" complex was also recorded in the cerebellum, but it developed there considerably later and its amplitude was much smaller than in the medulla. This complex was often observed in the electro-

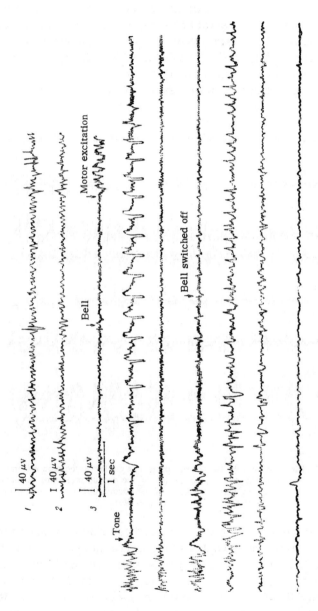

Fig. 46A. Electrogram of the action potentials of the brain of a rat during an epileptiform reaction. 1: Medulla oblongata; 2: hippocampus; 3: cerebellum (vermis).

encephalogram of the midbrain and hypothalamus, although it was much weaker here. The pathological activity at different stages of the fit was also recorded in animals rendered completely immobile with curare. Obvious epileptoid changes were found in the electro-encephalograms of certain divisions of the rhinencephalon (the hippocampus and the pyriform lobes), which, in the course of the systematic exposure of the animal to the sound stimulus, became more and more distinct from one experiment to the next.

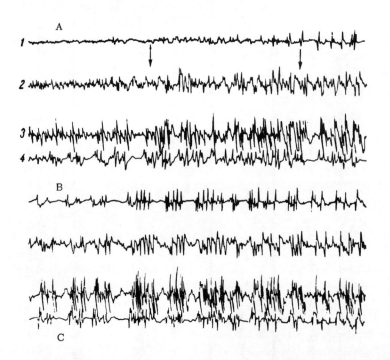

Fig. 46B. Electrogram of the brain of a rat during myoclonic spasms. 1: Motor region of the cortex; 2: region of the corpus striatum; 3: inner corpus geniculatum; 4: auditory region of the cortex. Calibration and notation of time as in Fig. 46A.

It may be postulated on the basis of these findings that impulses of excitation from the sound stimulus, reaching the auditory and vestibular nuclei in the medulla, create there a focus of pathological excitation, which from thence spreads upward, mainly by nonspecific paths. During the motor and clonic stages of the fit, the excitation radiates widely throughout the brain. During the tonic stage of the fit the excitation is concentrated in the region of the primary focus (the anterior division of the medulla oblongata).

Marked changes in the bioelectrical activity of the brain were found during the myoclonic spasms. Under these circumstances epileptoid discharges of high amplitude (500–800 μv) were observed. It must be pointed out that although the pathological activity occurred in all the divisions of the brain tested, the earliest and most obvious changes were found in the region of the subcortical centers of the auditory analyzer. In the cortex these impulses appeared much later (Semiokhina, 1958) (Fig. 46B).

It may be assumed from these results that the primary focus of excitation in myoclonic spasms is created in the auditory subcortical centers. From here the impulses of excitation spread to the motor area of the cortex, without which this particular pathological motor reaction, obviously, could not take place. This is in full agreement with the results of the previous experiments on extirpation of the cerebral cortex.

To sum up this chapter we shall return to our initial statements. These investigations showed that overloading the nervous system with afferent impulses, especially in individuals with increased excitability and weakness of inhibition, may cause a marked and prolonged excitation of the brain, which in turn often terminates in a variety of serious conditions: convulsive fits, myoclonus, a cataleptoid state, and, finally, acute circulatory disturbances. The latter may terminate in death from hemorrhage (usually into the brain).

The importance of inertia of excitation during the development of many pathological conditions in man has been pointed out by Davidenkov (1947).

The electrophysiological investigations and the decortication experiments described above give grounds for the belief that the reticular formation of the brainstem may play an essential role in the production and maintenance of the state of excitation caused by the exteroceptive stimulus in the pathological condition which we are studying.

The investigations of Moruzzi and Magoun (1949) and of Jasper (1949) showed that stimulation of the reticular formation of the brainstem and thalamus causes a considerable general excitation of the cerebral cortex. The excited state of the cortex, when due to stimulation of the reticular formation, is characterized by a longer duration than the same effect when obtained by stimulation of specific pathways. In addition to the general stimulating influence on the cerebral cortex, the reticular formation activates the vegetative functions, modifying, for example, the cutaneous galvanic reflex

(Wang, Stein, and Brown, 1956) and the vascular tone (Bonvallet, Dell, and Hiebel, 1954).

The probable role of the reticular formation in the maintenance of prolonged excitation and in the production of vegetative changes is also indicated by the morphological peculiarities of its structure: on the one hand, certain visceral functions of the body, such as the respiratory and vasomotor functions, are effected by groups of neurons in the reticular formation of the medulla itself; on the other hand, the neurons of the reticular formation have numerous branches, by means of which a great variety of connections may be made with other neurons, both in the reticular formation itself and beyond its limits (see Scheibel and Scheibel, 1958; Polyakov, 1959), forming a closed circle of return connections, thereby enabling the prolonged circulation of impulses derived from specific pathways.

It may therefore be postulated from these findings that when the nervous system is overloaded with afferent impulses, hyperactivation of the reticular formation takes place, which leads to a state of increased stationary excitation of both the subcortical formations of the diencephalon (the hypothalamus) and the telencephalon, and of the cerebral cortex; it causes prolonged vegetative disturbances, which may even culminate in death. This hyperactivizing influence of the reticular formation may evidently be exerted only when the inhibitory barriers of the nervous system are weakened, at the level of both the cortex and the subcortical and brainstem divisions of the brain.

This investigation shows the care with which all the factors leading to excitation of the brain and weakening of the inhibitory functions of the nervous system must be considered; not only may they cause severe functional disturbances of nervous activity, but they may also terminate in death.

CHAPTER 5

THE EXPERIMENTAL STUDY OF THE RUDIMENTS OF RATIONAL BEHAVIOR IN ANIMALS (EXTRAPOLATION REFLEXES)

During the last years of his life, Pavlov made repeated attempts to classify the associations, which he regarded as the basic elements of thought. At his "Wednesday meeting" of December 5, 1934, he said: "The whole art of teaching consists of the formation of temporary connections, which are the basis of thought, reflection, and knowledge. The fundamental factor, then, is association, thinking, a fact which some psychologists have known for a long time and have correctly used in formulating their opinions."* Later on he said: "In our experiments on artificial food-conditioned reflexes, when the connections are established these reflexes, having the meaning of food signals and of signals changing in accordance with the experimental conditions, have a dual character: they are signals and they are strictly temporary. When we are concerned with Thorndyke's experiments, these connections are then more permanent."† In these statements of Pavlov we see an attempt to classify associations into more particular cases.

A year later Pavlov made more detailed pronouncements on the classification of associations. He said: " . . . the term 'associations' is a generic concept, i.e., a combination of what was previously separated, a unification, a generalization of two points in a functional relationship, their fusion into a single association; but the term 'conditioned reflex' is a specific concept. This, of course, is also a combination of two points which formerly were not combined, but it is a special case of such a combination, having a particular biological significance. In the case of the conditioned reflex you have the essential constant features of a known object (food, an enemy, etc.), replaced by temporary signals. This is a special case of the application of an association.

"On the other hand, when two phenomena are connected by reason of the fact that they act simultaneously on the nervous system, phenomena are connected which are, in fact, permanently

*I. P. Pavlov, op. cit., Vol. 2, p. 580.
†Ibid, p. 581.

linked. It is this other form of association which is the basis of knowledge, the basis of the fundamental scientific principle of causality. This other form of association, perhaps no less and probably more important than conditioned reflexes, is a signal connection.

"Finally, in the simple case (which might be called artificial, accidental, nonessential, unimportant) when, for example, two sounds are connected psychologically, having nothing in common, they are connected solely by the fact that one follows the other, and when at last they are connected, one causes the other.

"These cases must, of course, be distinguished. They are all specific cases and specific concepts, but the associative connection is a generic concept."*

From Pavlov's statements it is quite clear that from the general concept of association he distinguished special cases of the formation of connections between phenomena. First, conditioned reflexes, formed as a result of the combination of two phenomena in the outside world which coincide in time, one of which must be of biological importance to the organism. In fact, however, the two phenomena are frequently not connected by a permanent causative connection, but merely coincide in time.

Second, artificial associations, formed as a result of the coincidence in time of two indifferent stimuli.

Third, associations formed between phenomena of the outside world permanently connected with each other. Pavlov cites an example of such an association: "When a monkey builds a tower in order to reach a fruit, this cannot be called a 'conditioned reflex.' This is a case of the formation of knowledge, of the establishment of a connection between objects. It is in a different category. This must be regarded as the beginning of the formation of knowledge, the forging of a permanent connection between objects—the factor which lies at the root of all scientific activity, of the laws of causation, and so on. I wished to draw attention to this matter. I have spoken about it, but from what I have said it is clear that no special attention has been paid to it. I shall now take another example."†

This important view of association, reflecting relationships of cause and effect existing between stimuli at a particular moment, has been little studied since Pavlov's death. In the present chapter we shall describe the results of our experimental studies of one case

*I. P. Pavlov, op. cit., Vol. 3, p. 262.
†ibid.

of this type of association, namely extrapolation reflexes* (Krushinskii, 1958a, 1958b, 1958c, 1959a).

We use the term extrapolation reflexes to describe the reaction of animals not merely to some direct stimulus, but also to the direction along which this stimulus proceeds as a result of its regular movement.

The ability to extrapolate, which is evidently brought about by means of rapidly forming associations between phenomena in the outside world and bearing a cause-and-effect relationship to each other, is in our opinion one of the most important criteria of rational activity. By reflecting cause-and-effect relationships between phenomena in the outside world, extrapolation reflexes ensure the adequate reaction of the animal to these relationships.

Attempts to establish criteria of rational activity have been made by several workers. Köhler (1930) concluded from his experiments on anthropoid apes that these animals could make "genuine decisions." The decision which had to be made was an unexpected one, not calling for any previous individual experience on the part of the animal. The main criterion of rational behavior is the solution of a problem taking account of the situation as a whole. "For this reason," he writes, "this sign, the appearance of a solution as a whole, to suit the requirements of the situation, must be accepted as a criterion of rational behavior." The criterion of rational behavior suggested by Köhler, however, cannot be used as a basis for the physiological analysis of the phenomenon which we are studying, for it is defined in very general terms.

Russell (1932, 1946) regards "insight" (a term introduced by Köhler) as the ability of animals, before carrying out any behavioral act, to make the decision to do so. Russell distinguishes such acts of behavior from behavior which the animal learns as a result of "trial and error."

Konorski (1950) also claims that animals have a special type of behavior, which is based upon reasoning or insight.

Under the meaning of "Intelligenz," Bierens de Haan (1931a, 1931b) understands the ability of an animal to retain traces of impressions received and to utilize its individually acquired experience during later life. On the basis of this experience the animal is able to understand phenomena arising by cause and effect (Kausalabläufe).

*Extrapolation means the determination of the pattern of change of a certain magnitude in the future on the basis of knowledge of the pattern of its change in the past.

Very similar ideas were expressed by Fischel (1949, 1953, 1956), who considers it possible to speak of insight (Verständnis) in cases when the animal is aware of the relationship existing between a particular action and the result to which it may lead. He is inclined to believe that association lies at the basis of insight. Fischel considers that only monkeys possess the ability to make a decision regarding the performance of a given action, and that for other animals preliminary individual experience is necessary. Fischel thus widens his definition of insight, which he regards as the ability to make a decision about what to do in the future.

Besides "insight" Thorpe (1958) distinguishes the adaptive response on the basis of insight (insight learning). By "insight" Thorpe implies the "organization of perception" or the establishment of relationships. So far as the second category of behavior is concerned, in Thorpe's opinion this is the spontaneous response reaction of the animal, without preliminary trial and error, or in other words the solution of a problem by means of the spontaneous adaptive reorganization of its previous experience.

Beritov (1947) raised the question of the presence in animals and man of another form of activity besides the reflex form of activity of the central nervous system, consisting of unconditioned and conditioned reflexes, namely "neuropsychic activity," lying at the basis of arbitrary movements. The principal behavioral acts which Beritov studied were certain very interesting but seldom analyzed responses of freely moving animals to the command, "To your place!" It was found that the rate at which the reflexes were established, the duration of retention of the stimulus trace, and the path along which the animals run to the place where the food is to be found, vary extraordinarily among the animals.

The results of this investigation are of great interest for the comparative physiological study of animal behavior. However, Beritov's views regarding the presence of "neuropsychic" activity in addition to reflex activity in animals have been repeatedly criticized in the physiological literature (Ivanov-Smolenskii, 1950; Kupalov, 1950; Voronin, 1951; and others).

Lodygina-Kots (1923, 1928, 1935, 1957, 1958) describes detailed investigations, lasting many years, of the formation of complex forms of constructive and tool-handling activity in anthropoid apes. These investigations revealed the distinctive features of that activity of the apes which is defined as the specific form of thinking (intellect) characteristic of anthropoids. This writer draws attention to the qualitative properties of the intellect of anthropoid apes in comparison to the human intellect. Anthropoid apes, in contrast to man, cannot operate mentally with pictorial images, so that the space-time connections which are formed in these animals do not change into cause-and-effect connections. The behavioral forms of anthropoids studied by Lodygina-Kots are so complicated and complex that it is difficult at present to analyze the physiological mechanisms on which they are based.

The study of complex forms of higher nervous activity and, in particular, of "arbitrary" movements in animals, has been pursued successfully in the laboratories of Kupalov (Kupalov, 1955, 1958; Alekseeva, 1956), Voronin (Rokotova, 1954; Napalkov, 1957, 1958), and others. In these investigations, complex systems of motor reflexes were formed in the animals, and the patterns of determination of chains of conditioned reflexes, defined as arbitrary movements, were studied.

The foregoing facts suggest that the study of complex forms of animal behavior, which may be regarded as the manifestation of elementary rational activity, has occupied the attention of several researchers. We consider that this question must be studied entirely within the framework of the physiological analysis of higher nervous activity. There is no doubt that the whole activity of the brain is reflex in nature, i.e., determined by external and internal influences, which are received by the animal's receptors.

The objective physiological analysis of complex forms of behavior of various animals, when carried out parallel with the study of the morphological structure of the brain, must undoubtedly provide abundant information on the formation of the rudiments of rational activity.

When selecting material in which to study extrapolation reflexes, we thought it desirable to begin our experiments with animals having relatively simple forms of higher nervous activity, for it could be expected that in such animals the reflexes to be studied (if they were present) would appear in their simplest form.

Convenient objects for this purpose were found to be birds and rabbits.* The birds which we studied in greatest detail were pigeons, ducks, fowls, and members of the crow family (crows, magpies, and rooks).†

---

*In order to exclude the possible role of olfaction in these experiments, in all the rabbits used, as a preliminary step the olfactory bulbs were destroyed.
†A few predatory birds were also studied (Krushinskii, 1958a, 1958c).

As a food stimulus we used millet for the fowls, hemp seed for the pigeons, bread and oats soaked in water for the ducks, meat and eggs for the crow family, meat for the predatory birds, and carrots and beets for the rabbits. The experiments were carried out in the laboratory. During the experiments the animals were able to move about the room freely.

The scheme of the experiments was as follows. The food stimulus (A) is moved in a straight line with constant velocity. The first section of its path of movement takes place in full view of the animal, which is able not only to see the food stimulus A but also to take food from it. After it has completed the part of its path in full view of the animal, the stimulus A is hidden behind the cover B.

Observations were made to see if the animals continued to search for the unconditioned stimulus after they ceased to perceive it with their receptor apparatus, and whether they were able to extrapolate the direction of its movement.

The animals were thus required to do the following: 1) as a result of the associations formed between A and B to continue to search for stimulus A around the obstacle B; 2) from its relationship to fixed points in space, to determine the change in the position (movement) of the point (of the stimulus) A; 3) having determined the unknown magnitude (the direction of movement of A) by the change in this relationship, to carry out the search in the direction of movement of A.

Concrete experiments were carried out in accordance with the general principles enunciated above. In the first variant of the experiment a specially constructed corridor was used.

The technique of the investigation was as follows. The feeding bowl containing food was moved along the track at a mean velocity of 8-10 cm/sec. The first 1.5 m of the track was open, so that the animal could walk behind the feeding bowl and eat food from it. The feeding bowl then entered the closed corridor (or tunnel). As soon as the food entered the corridor, the entrance was covered by a flap and the animal could no longer see the food as it moved along the corridor. The corridor consisted of two halves, each 1.5 m in length. In some variants of the experiment a space of 3-5 cm was left between the two halves of the corridor, through which the animal could see the feeding bowl with the food at the time when it passed the space.

The following were observed during the experiment: 1) whether the animal looked for food at the place where it had disappeared, or in the direction of its movement along the corridor; 2) for how long and how far it searched for food along the corridor.

A control experiment was first carried out. An empty feeding bowl moved along the corridor. A difference in the animal's behavior during movement of the empty feeding bowl and of the feeding bowl with food would show whether the animal was, in fact, seeking food after it had disappeared, or whether it was displaying an orienting reaction to the slight sound of the feeding bowl as it moved along the corridor.

In order to ascertain whether the noise of the feeding bowl containing food as it moved along the corridor acted as a guide to direct the animal in its search, control experiments were carried out in which the feeding bowl was brought to a halt as soon as it was concealed in the corridor. These experiments showed that the noise made by the moving feeding bowl did not act as a guide to the animal during its search for the food as it moved along the corridor.

Fig. 47. Scheme of the experiment with the pigeons: 1—direction of movement of the food; 2—of the bird. The broken line designates the movement of the bird before the disappearance of the food in the corridor; the continuous line—after the disappearance.

Experiments with pigeons. The results of the experiments with pigeons were remarkably consistent. In all 14 experiments the pigeons followed behind the moving feeding bowl and took food from it, but as soon as the bowl disappeared in the corridor most of the birds at once turned back from the corridor and walked along the track along which the food had just moved.

Pigeon No. 5. Experiment on May 14, 14 hr 22 min. Velocity of movement of feeding bowl 10 cm/sec. The feeding bowl moved along the track 1.5 m, after which it was concealed in the corridor. The pigeon followed the feeding bowl and took food from it. As soon as the bowl disappeared in the corridor, the pigeon at once turned away from the corridor, crossed over to the other side of the track, and went along it in the other direction, opposite to the movement of the food (Fig. 47*).

Only in one experiment, after disappearance of the bowl with the food, did the pigeon (No. 1) walk to one side of the corridor and wait there, pluming itself. In two other experiments the pigeons

*The schemes illustrate the character of the animals' movement, but are not drawn accurately to scale.

(Nos. 2 and 3) made very short (3 and 1 sec) pauses near the beginning of the corridor after the food had disappeared inside.

Pigeon No. 2. Experiment on April 14, 17 hr 15 min. Velocity of movement of feeding bowl 10 cm/sec. The feeding bowl moved along the track for 1.5 m, after which it was concealed in the corridor. The pigeon followed the feeding bowl and took food from it. As soon as the bowl disappeared in the corridor, the bird took a few steps away from the beginning of the corridor, turned its head toward the corridor for 3 seconds and looked at the place from which the food had disappeared, after which it turned back along the track and then made off at an angle of 45°.

The experiments showed that, after the disappearance of the food in the corridor, the pigeons made no attempt to look for it. Furthermore, the pigeons usually turned back along the track immediately after the disappearance of the food in the corridor, in the opposite direction to that taken by the feeding bowl from which they had only just taken food.

Table 39. Duration of Stay of Fowls near the Beginning of the Corridor during Movement of the Feeding Bowl With and Without Food (Results of the First Experiment)

Serial No.	Name or number	Duration of search during movement of feeding bowl (in sec)		Serial No.	Name or number	Duration of search during movement of feeding bowl (in sec)	
		with food	empty			with food	empty
1	Levyi	70	10	13	Sova	2	0
2	Pravyi	43	4	14	1715	33	0
3	Bandit	0	0	15	Chernushka	6	7
4	Kryuchok	7	0	16	Pestrushka	0	0
5	Pryamoi	16	0	17	Boevik	0	0
6	1714	0	0	18	Ryzhik	4	0
7	1728	0	0	19	Yurkii	12	0
8	1724	18	0	20	Krasavets	4	0
9	1702	10	0	21	Spornyi	12	0
10	1713	30	0	22	Khvastun	5	0
11	Rogul'ka	3	0	23	Tochka	0	0
12	1742	18	0	24	Mokhnatka	7	0

Experiments with ducks. Experiments with four ducks showed that they also made no attempt to seek the food along the corridor. In some experiments, however, they were observed to stay (for 7–22 seconds) near the place from which the food had disappeared. Under these circumstances, characteristic movements of the head and beak were observed toward the place from which the food had disappeared. In the ducks, therefore, although no extrapolation-reflex acts of movement along the corridor could be seen, the track of the stimulus was preserved longer after the cessation of its action in these birds than in the pigeons.

Experiments with fowls. These showed that fowls differed from both pigeons and ducks in their reaction to the disappearance of the food in the corridor. The principal feature characterizing the behavior of the majority of fowls after the disappearance of the food was that they searched around the place of disappearance of the food in the corridor. In order to verify that the fowls' lingering at the beginning of the corridor was not fortuitous, control experiments were carried out in which the empty feeding bowl was moved (Table 39).

By way of example of the search for the food as it disappeared in the corridor we may describe an experiment with a cock with the name of Levyi.

Experiment on April 11. Velocity of movement of feeding bowl 10 cm/sec. Open path of movement of feeding bowl 1.5 m.

14 hr 25 min. The cock followed the empty feeding bowl and examined it. After the bowl had disappeared into the corridor, the bird stood still for 10 seconds near the beginning of the corridor and, going to one side of the corridor, it went along it and then turned back.

14 hr 30 min. Feeding bowl with food. The cock followed the bowl and took food from it eagerly. When the food had disappeared in the corridor the bird immediately flew onto the top of the corridor and began to scratch its edge with its claws, above the place from which the food had vanished from sight. The bird scratched and circled at the beginning of the corridor for 70 seconds, after which it jumped from the corridor and went away (Fig. 48).

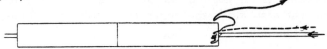

Fig. 48. Scheme of the experiment with the cock (Legend as in Fig. 47).

In some birds of this species searching movements were observed not only at the beginning of the corridor but also in the direction of movement of the food as it disappeared in the corridor. By way of example we may cite an experiment with fowl No. 1724.

Experiment on October 19. 18 hr 20 min. Velocity of movement of feeding bowl 9 cm/sec. Open path of movement of feeding bowl 1.5 m. The fowl did not react to the moving, empty feeding bowl. It walked about near the beginning of the corridor (Fig. 49).

18 hr 25 min. The fowl followed the feeding bowl containing food and ate from it. When the bowl disappeared in the corridor the bird stretched out its neck, went along the corridor for a distance of 0.5 m, and all the time tried to glance through the top of the corridor; returning to the beginning of the corridor, it crossed over to the other side and moved along it for 25–30 cm, also trying the whole time to peer through the top of the corridor; it again returned to the beginning of the corridor, went away from it, and turned back along the track. The search for food had lasted 18 seconds (Fig. 50).

This experiment shows that, after the disappearance of the food in the corridor, some fowls not only seek it at the beginning of the corridor, but make short extrapolation-reflex movements along the corridor (for a distance of 0.5 m). The most characteristic

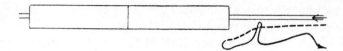

Fig. 49. Scheme of the experiment with the fowl. Reactions to
movement of the empty feeding bowl. (Legend as in Fig. 47.)

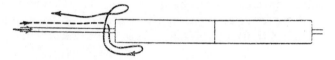

Fig. 50. Scheme of the experiment with the fowl. Reactions to the
disappearance of the feeding bowl with food in the corridor. (Legend
as in Fig. 47.)

feature of the behavior of the fowls, however, is their continued
search for food at the place from which it disappeared (the beginning
of the corridor).

Experiments with birds of the crow family. These experiments showed
that, by the character of their search for the food as it disappeared
in the corridor, these birds differed from both pigeons and fowls.
All eight birds investigated looked for food very actively and for
a considerable time, not only at the place from which it had disappeared, but also along the corridor. Hardly any seeking for the
empty bowl took place (Table 40).

Table 40. Time and Distance of Seeking for Food along the Corridor by Birds of the Crow Family during Movement of the Feeding
Bowl With and Without Food*

Serial No.	Name	Species	Search during movement of feeding bowl			
			with food		without food	
			time (in sec)	distance (in cm)	time (in sec)	distance (in cm)
1	Pava	Crow	>50	300	0	0
2	Mashka	Crow	45	150	0	0
3	Varya	Crow	47	240	0	0
4	Varyag	Crow	75	120	2	10
5	Kralya	Crow	51	230	0	0
6	Zhulya	Magpie	>15	150	0	0
7	Bezymyanka	Magpie	80	120	10†	0
8	Artemida	Magpie	50	150	0	0

*Results of first experiment.
†Length of stay near beginning of corridor.

As an example of a search pattern for the disappearing food, we may describe the experiment with the crow given the name of Varya.

Experiment on October 1. Velocity of movement of feeding bowl 8 cm/sec. Open path of movement of bowl 1.5 m.

15 hr 20 min. The crow ran after the feeding bowl, which had a piece of meat fixed to it, and pecked it. As soon as the bowl disappeared in the corridor, the bird ran alongside the corridor for a distance of 240 cm, turned back to the beginning of the corridor but, before reaching it, turned again and went forward in the direction of movement of the food. After 47 seconds the search for the food was abandoned and the crow went away from the corridor (Fig. 51).

15 hr 25 min. The crow did not react at all to the moving empty feeding bowl, but turned away from it.

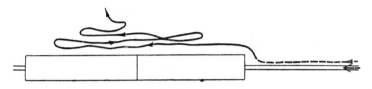

Fig. 51. Scheme of the experiment with the crow. (Legend as in Fig. 47.)

Some very instructive experiments were carried out with birds of the crow family, in which the two halves of the corridor were separated by a small gap, through which the birds could see the food as it passed that point. In two cases the birds, when they saw the food passing the gap, immediately ran along the second half of the corridor to the end, where they waited.

Experiment No. 1 with the crow Zhulya, February 11. Velocity of movement of feeding bowl 10 cm/sec. Gap between two halves of corridor 4 cm.

12 hr 45 min. As soon as the bowl, containing meat, began to move, the crow ran after the food and pecked it. When the food disappeared in the corridor, Zhulya ran alongside the corridor for a distance of 60–70 cm, turned back to the beginning of the corridor and then quickly ran to the gap, peered into it, made a short jump backward, and again ran to the gap. At this moment the food was passing the gap. Zhulya at once rushed forward along the second half of the corridor for 50–60 cm, turned back quickly to the gap, peered into it and rapidly ran to the end of the corridor, where it waited; the bird's head was turned toward the end of the corridor; it was tense and on the alert. It stood at the end of the corridor until the food reappeared;* it then pecked the food (Fig. 52).

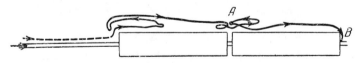

Fig. 52. Scheme of the experiment with the crow: A—position of bird at the moment when the food moved past the gap between the two halves of the corridor; B—position of the crow when the food emerged from the corridor.

*The bird did not see the food moving in the corridor, because the outlet of the corridor was covered by a flap.

In this experiment in order to reach the food Zhulya made two extrapolation-reflex excursions: along the first half to the gap between the two halves of the corridor, and along the second half of the corridor after having seen the food as it moved past the gap.

A similar result was obtained in the experiment with the crow Artemida, which made a precisely similar extrapolation-reflex excursion alongside the first half of the corridor and, when it saw the food passing the gap, ran on to the end of the corridor and waited there in a tense, expectant attitude.

The short individual experiment in the course of which the bird established an association between the direction of movement of the stimulus and the environment in which this movement took place was evidently sufficient to make it run the whole length of the first half of the corridor. The sight of the continuing movement of the food past the gap was further reinforcement of this association, and this led to the excursion of the bird along the second half of the corridor to its end, where it "awaited" the food.

These experiments showed the vast difference in the character of the search for the food disappearing in the corridor made by all eight birds of the crow family that were tested, in comparison to that of the fowls and pigeons.

Experiments with rabbits. These experiments showed that in the character of their seeking for the food which had disappeared in the corridor, rabbits behaved like fowls (Table 41). From both the first and subsequent experiments it may be concluded that rabbits continue to seek the food for a short time at the place of its disappearance, and in one case the animal searched alongside the corridor. As an illustration we cite the result of the experiment with the rabbit Kol'ka.

Experiment with the rabbit Kol'ka. 15 hr 25 min. Feeding bowl with carrot moved along the track. The rabbit followed behind the bowl and gnawed the carrot. When the bowl disappeared in the corridor, the rabbit continued to move up and down persistently by the beginning of the corridor, jumped over the track a few times, ran along the corridor on both sides for its full length, then ran alongside the corridor as far as the gap, into which it peered (at this moment the food had already passed the gap), and after 40 seconds ran away from the corridor.

In most experiments, however, after the disappearance of the food the rabbits merely stayed at the beginning of the corridor.

Experiment with the rabbit Serka, October 28. 14 hr 33 min. The rabbit ran after the moving feeding bowl containing carrot, and gnawed it. When the bowl disappeared in the corridor, Serka sat for 7 seconds by the entrance to the corridor, after which it ran away and gave no further reaction to the situation.

In all the subsequent experiments the same pattern was observed: the rabbits either waited or, in some cases, they tried to

Table 41. Duration and Distance of
Search for Food Alongside the
Corridor by Rabbits*

Name	Search for food during movement of feeding bowl	
	Time (in sec)	Distance (in cm)
Kol'ka.......	40	150
Visloushka...	22†	0
Serka	7†	0
Karlik.......	5†	0
Pyshka	12†	0

*Results of first experiment.
†Duration of stay near beginning of corridor.

seek food at the beginning of the corridor, or they made short
excursions alongside the corridor to seek the food.

To sum up, the following conclusion may be drawn. A great
difference is observed in the degree of expression of extrapolation

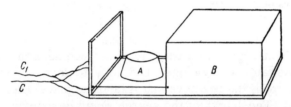

Fig. 53. Scheme of the platform with the moving box (B)
by the feeding bowl (A). C—cord which moves the platform;
C_1—cord by which the box is moved over the feeding bowl.

reflexes during the search for the food as it moved and disappeared
in the corridor. In pigeons and ducks these reflexes were not found
(the ducks merely waited a short time at the beginning of the corridor). Fowls and rabbits looked for the vanished food in the place
where they had last seen it, and in some cases this search was also
continued alongside the first part of the corridor. In the birds of
the crow family the ability to extrapolate the direction of movement
of the food disappearing in the corridor was very highly developed.

The next variant of the experiment utilized a moving platform.
The experiment was conducted as follows. On a platform, measuring
18.5×37.5 cm, was placed a box which moved along grooves and
which could be brought up against the front wall of the platform,

whereupon the food placed on the platform could not be seen by the animal. The animal in this experiment ate from a feeding bowl placed on the moving platform (Fig. 53). After the platform had moved forward a distance of 1.5 m, the food was covered by the box, and the platform either stopped at once or continued to move. The velocity of the movement of the platform was 4–12 cm/sec.

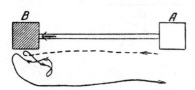

Fig. 54. Scheme of the experiment with the moving platform and the pigeon. The unshaded square represents the feeding bowl with uncovered food; the shaded square the bowl with the food covered. A—original position of the feeding bowl; B—the place at which the food was covered. The broken line denotes the movement of the animal before the food was covered; the continuous line, after the food was covered.

We investigated whether rapidly forming associations are developed in the animals between the food and the box which covered it, i.e., whether any search for food took place around the platform with the concealing box. We accordingly determined the difference in the search for food in the following two cases: a) after the food had been covered by the box, the platform was immediately halted; b) after the food had been covered by the box, the platform continued to move in the same direction.

In the first case, in order to seek food near the stationary platform, the animal evidently had only to form an association between the food, the box covering it, and the place where the food was covered. The length of the search would be determined by the time that this association lasted.

If the platform continued to move after the food had been covered, the direction of movement of the platform was added to the composition of the developing association. In this case the animal had to carry out an elementary act of extrapolation; it had to seek food not at the place where it had been covered by the box, but around the platform which continued to move.

Experiments with pigeons. When the food was covered and the platform was simultaneously halted, the pigeons either at once went away from the platform or (in one half of the experiments) they stayed for a short time (4–8 seconds) near the stationary platform. Only in one case did the pigeon go away for 1 min 20 sec and then return to the stationary platform.

In the experiments in which the food was covered by the box and the platform subsequently continued to move, in most cases the pigeons at once went away from the moving platform without giving any further reaction to it. In only one of the 19 experiments did the pigeon take a few steps behind the food, as it remained covered on the moving platform, and then go away; another pigeon, after leaving the moving platform on which the food was hidden by the box, made two turns around the laboratory, and then came up to the moving platform again, only to leave it immediately.

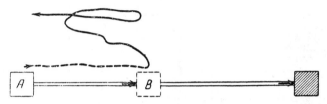

Fig. 55. Scheme of the experiment with the moving platform and the pigeon. (Legend as in Fig. 54.)

By way of example we describe the results of the experiments with one pigeon.

Experiment on April 2. 13 hr 15 min. The platform on which stood the feeding bowl containing the food moved at a velocity of 5 cm/sec. The pigeon followed behind the platform and pecked at the food in the bowl. After the platform had moved through a distance of 1.5 m, the box was moved over the food, and the platform brought to a halt. The pigeon went a few steps away, came back to the platform again, stretched out its neck, seemed to be trying to look through the box, and then went away from the platform. The reaction to the stationary platform on which the food was covered lasted 8 seconds (Fig. 54).

13 hr 35 min. The platform moved at a velocity of 4.5 cm/sec. The pigeon followed behind the platform and pecked from the feeding bowl on the platform. After the platform had moved through a distance of 1.5 m, the food was covered by the box, while the platform continued to move. As soon as the feeding bowl was covered, the pigeon immediately went away from the platform (Fig. 55).

These experiments thus showed that after the food was covered by the box, irrespective of whether the platform stopped or continued to move, the pigeons as a rule ceased to react to it after a few seconds.

Experiments with ducks. The behavior of the ducks in these experiments contrasted sharply with that of the pigeons. If the platform was halted after the food was covered by the box, the ducks continued to react energetically to the stationary platform: they walked around it and tapped on the box with their beak. In one case this search for the food continued for 70 seconds.

If, however, the platform continued to move after the food was covered by the box, the ducks at once ceased to react to the moving platform. They stayed at the place where the food was covered for 2–3 seconds, and then walked away. Only in one case did a drake take a few paces behind the moving platform.

Table 42. Reaction of Fowls to Food Remaining Stationary or Continuing to Move after Being Covered*

Name or number of bird	Stationary platform	Platform continuing to move	
	Duration of reaction (in sec)	Duration of reaction (in sec)	Movement behind platform (in cm)
Bandit.	17	0	0
Kroshka.	32	1	10–12
Zheltushka	30	0	0
Pravyi	26	16	125†
Kryuchok.	30	0	0
Khvastun	55	0	0
1714	40	5	40
1728	55	8‡	5–6
1724	65	3‡	0
Chernushka	38	8‡	0
Golyshka	44	8‡	0
Pestrushka.	37	9‡	0
Mokhnatka	57	10	100
Rogul'ka	16	0	0
Tochka	18	0	0
Ryzhik.	12	5‡	0
Yurkii.	14	0	0
Krasavets	15	6‡	0
Spornyi	80	3‡	0
Sova	13	10‡	0
Boevik.	8	5‡	0

*Results of the first test with each bird.
†Control experiments carried out with these cocks showed that they followed behind various moving objects, displaying signs of an orienting reaction. It is probable that in the experiment described the movement behind the covered and moving food is also associated with the orienting reflex of the cock to a moving object.
‡When the food on the moving platform was covered by the box, the fowls did not move, but continued to remain where they were.

Experiments with fowls. The results of these experiments are shown in Table 42.

It may be seen from Table 42 that when the food was covered and the platform simultaneously halted, all the fowls reacted for 8–80 seconds to the box which covered the food. They actively searched for food on the stationary platform with the box (Fig. 56). When the platform continued to move after the food had been covered by the box, the fowls either went away from it immediately or remained for a further 6–9 seconds at the place where the food had been covered, sometimes making scratching movements with their claws. Only a very few followed behind the covered food on the

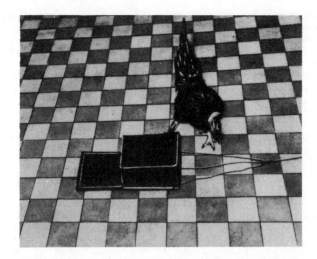

Fig. 56. Cock looking for food near the stationary platform.

moving platform. As an illustration of a typical case we may mention
the experiments with the hen Chernushka.

Experiment on December 20. 17 hr 33 min. Platform on which stood an empty feeding
bowl moved at a velocity of 11 cm/sec. After the platform had moved through a distance
of 1.5 m, the feeding bowl was covered by the box, but the platform moved a further 2.5 m.
As soon as the food was covered, the bird immediately stopped; it stood for 8 seconds
and then walked back (Fig. 57).

17 hr 38 min. The platform on which stood the feeding bowl containing the food moved
at a velocity of 9 cm/sec. The hen followed behind the platform and pecked the food from
the feeding bowl. After the platform had moved through a distance of 1.5 m, the feeding
bowl was covered by the box, and the platform moved a further 2.5 m. As soon as the food
was covered the bird stopped, stood still for 8 seconds, and then walked away.

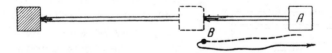

Fig. 57. Scheme of the experiment with the moving platform and
the hen. The black circle indicates the place where the bird
stopped after the food was covered. (Legend as in Fig. 54.)

17 hr 45 min. The platform on which stood the feeding bowl containing food moved at
a velocity of 8 cm/sec. The hen followed behind the platform and pecked the food from
the feeding bowl. After the platform had moved through a distance of 1.5 m, the bowl was
covered by the box moving over it, and was simultaneously halted. The hen walked around
the platform, and once it tapped on the box. After 38 seconds it went away (Fig. 58).

As an example of the movement of the birds behind the moving
platform, we may cite the experiment with hen No. 1714.

Experiment on December 10. 17 hr 48 min. The platform, on which stood the feeding bowl containing food, moved at a velocity of 8 cm/sec; the length of its track was 4 m. After the platform had moved 1.5 m, the box was pulled over the food; after the food had been covered the hen continued for 5 seconds to follow behind the moving platform, for a distance of 40 cm.

These experiments thus showed that all the hens searched for the covered and stationary food, but only a few of them reacted to the covered but still moving food.

Table 43. Reaction of Birds of the Crow Family to Food Continuing to Move or Remaining Stationary after Being Covered*

Name	Species	Stationary platform Duration of reaction (in sec)	Platform continuing to move Duration of reaction (in sec)	Movement behind platform (in cm)
Pava.............	Crow	45†	>29	>250
Mashka..........	Crow	–	>10	> 80
Varya...........	Crow	124†	8	90
Varyag..........	Crow	53	35	250
Zhulya..........	Magpie	–	10	100
Bezymyanka.......	Magpie	30	38	>250
Artemka.........	Magpie	88	30	>250

*Results of the first test of each bird.
†The crow was able to move the box and reach the food.

Experiments with birds of the crow family. Investigations carried out with birds of the crow family showed that they sought food near the platform with food covered by the box, whether it was stationary or continued to move (Table 43).

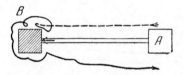

It may be seen from Table 43 that all the crows tested reacted to the platform on which the food was covered, irrespective of whether it was stationary or continued to move.

Fig. 58. Scheme of the experiment with the moving platform and the hen. (Legend as in Fig. 54.)

In all cases the reaction to the covered food was well marked. The birds diligently strove to peer through or to poke under the box covering the food. In order to verify that this was not an orienting reaction to a new moving object, the experiment began in all cases with movement of the empty platform, without food. As an illustration of the behavior of the bird during the experiment, we may cite the example of the magpie Bezymyanka.

Experiment on April 28. 14 hr 30 min. Empty platform covered by box moved at a velocity of 10 cm/sec. At the moment of movement of the platform the magpie was near it. It did not react to the platform, but went away from it, and did not show the slightest reaction throughout the whole time of movement of the platform.

14 hr 47 min. The platform on which stood the feeding bowl containing food moved at a velocity of 5 cm/sec. The magpie pecked at the meat, but showed signs of fear of the moving platform. After the platform had moved 1.5 m, the food was covered by the box and simultaneously halted. In the course of 30 seconds the magpie twice ran the full length of the platform on both sides and tried to peer beneath the box. It then ran away and gave no further reaction to the platform (Fig. 59).

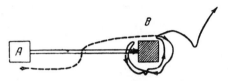

Fig. 59. Scheme of the experiment with the moving platform and the magpie. (Legend as in Fig. 54.)

14 hr 55 min. The platform on which stood the food moved at a velocity of 6.5 cm/sec. The magpie ran and pecked at the meat. After the platform had moved 1.5 m, the food was covered by moving the box over it, and the platform continued to move for a further 2.5 m. After the food had been covered, the magpie continued to run behind the platform, first on one side and then on the other, and all the time trying to peer under the box. Once it tapped on the box. It ran all the time that the platform was moving; when it came to a halt the bird ran the full length of the platform on both sides, after which it went away (Fig. 60).

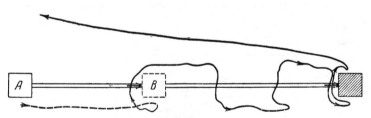

Fig. 60. Scheme of the experiment with the moving platform and the magpie. (Legend as in Fig. 54.)

The investigations with crows and magpies showed that their behavior in the experiment described differed considerably from that of the pigeons and fowls. Their reaction to the covered food was distinct, irrespective of whether the food was stationary after it had been covered, or continued to move.

Experiments with rabbits. These experiments showed that rabbits continued to search for covered food irrespective of whether the platform on which the food was placed was stationary or continued to move after the cover was applied (Table 44).

Table 44. Reaction of Rabbits to Food Continuing to
Move or Remaining Stationary after Being Covered*

Name	Stationary platform	Platform continuing to move	
	Duration of reaction (in sec)	Duration of reaction (in sec)	Movement after platform (in cm)
Kol'ka	14	33	170
Visloushka	90	25	150
Serka.	90	20	150
Pyshka	47	30	150

*Results of the first test of each animal.

Experiment with the rabbit Serka. December 20, 11 hr 30 min. The empty platform moved at a velocity of 11 cm/sec. The platform began to move when the rabbit was near by. The animal at once ran away from the platform and did not react to it.

11 hr 34 min. The platform on which stood the feeding bowl containing food (carrot) moved at a velocity of 8 cm/sec; length of track 4 m. The rabbit followed the platform and gnawed the carrot. After the platform had moved 1.5 m, the food was covered with the box. The animal continued to follow behind the platform, and at times thrust with its muzzle at the box.

11 hr 47 min. The platform on which was standing the feeding bowl containing food moved at a velocity of 8.5 cm/sec. The rabbit followed behind the platform and gnawed the carrot. After the platform had covered a distance of 1.5 m, the food was covered and the platform stopped. For 35 seconds the rabbit continuously moved up and down by the platform, trying to poke its head under the box, and having run around the platform on every side, it ran away to a distance of 2.0–2.5 m; after 65 seconds it again ran toward the platform and along its whole length; after 90 seconds it ran away from the platform, to which it did not again return (Fig. 61).

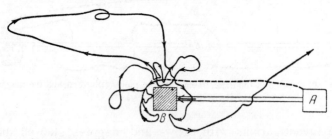

Fig. 61. Scheme of the experiment with the moving platform and the
rabbit. (Legend as in Fig. 54.)

The experiments described show that rabbits possess the ability to seek covered food not only when this is stationary but also if it continues to move.

These investigations (the experiments with the corridor and the platform) showed that animals possess the ability to develop rapidly forming associations between unconditioned reflex stimuli and obstacles preventing these stimuli from being received by their receptor apparatus. The indifferent stimulus (the obstacle) B, after it has covered the unconditioned reflex stimulus A, ceases to be indifferent. AB* in some species of animals elicits a search reaction.

In pigeons the association between A and B is very unstable; after the disappearance of A as a result of the obstacle B, these birds almost immediately cease to seek the unconditioned reflex stimulus. In ducks an association is formed between A and B (in the experiments with the platform); they search around the stationary platform with the covered food. In fowls an association is formed between A and B. For some tens of seconds they search for food at the place where it stopped and disappeared behind the obstacle B; in some individuals a search for food was made not only at the place where it disappeared, but also in the direction of its movement. This search in the direction of movement of the food, however, was usually brief.

Similar results were also obtained in rabbits which, besides searching at the place where the food disappeared, evidently possessed the ability to seek food also in the direction of its movement.

A definite ability to form associations with the direction of movement of the stimulus was observed in birds of the crow family, which searched for food not only at the place where it had disappeared, but also in the direction of its movement (the corridor experiment) and around the platform which continued to move after the food standing on it had been covered by the box. The associations formed in these birds thus included evidently not only the obstacle B but also the direction of movement of the unconditioned reflex stimulus. Extrapolation reflexes are of great importance in the behavior of these birds.

The results of these experiments, showing the presence in animals of extrapolation reflexes, evidently formed on the basis of rapidly established associations, justified further investigations in this direction.

In subsequent experiments we studied the importance of extrapolation reflexes in the search for a stimulus not only in the case of an unrehearsed task, when the animal's reaction had not previously been reinforced by the unconditioned reflex stimulus, but also in experiments in which a conditioned reflex could be formed in an animal as a result of the reinforcement of correct solutions.

*AB implies that the unconditioned reflex stimulus A is not perceived, as a result of the action of the obstacle B.

The experiments were carried out as follows. The animal fed itself through a vertical gap in a screen from one of two feeding bowls standing side by side: one containing food (experimental), the other empty (control). The gap was in the middle of the screen. The length of the screen was 2 m, the height 75 cm; the two feeding bowls stood on small platforms which could be moved along a track laid alongside the screen. The width of the gap, which could be adjusted to suit the size of the animal being studied, was controlled in such a way that the animal could put its head through but not its body.

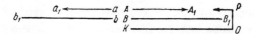

Fig. 62. Scheme of the experiment with the screen. b_1B_1—screen; bB—gap in screen; a—initial position of empty feeding bowl. A—of bowl containing food; a_1—position of empty bowl after movement; A_1—position of bowl containing food after movement; K—initial position of animal; KOP—movement of animal to food.

After the animal has eaten food for a few seconds from the feeding bowl, the two bowls (the one with food and the empty one) begin to move at the same velocity in opposite directions on the track alongside the screen, so that the animal is unable to reach

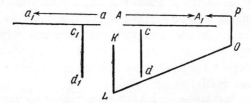

Fig. 63. Scheme of the experiment with the screen and attachment. (Legend—see Fig. 62 and the text).

the food. In order to obtain food, the animal has to run the length of the screen in the direction of movement of the bowl containing food. If the animal does this correctly it can obtain food from the bowl behind the screen, but if it runs to the opposite end of the screen, to the empty feeding bowl, the bowl containing food is taken away, and this excursion is not reinforced by food (Fig. 62).

In this experiment, in order to reach the feeding bowl containing food, the animal must follow the path KOP (in the position P it can

see the bowl of food). The initial direction of movement of the
animal (KO) is parallel to the direction of the movement of the
bowl (AA_1). The animal is allowed to seek food for 1.0-1.5 minutes
after the beginning of the movement of the feeding bowl.

If it runs in the correct
direction, i.e., in the direction
of the food, the task is made
more complicated. The attachment $c_1 d_1 dc$ (Fig. 63) is joined
to the screen near the gap. In
this variant of the experiment,
in order to reach the food, the
animal must first turn through
180°, and then move in the direction KL, perpendicular to
the direction of movement of
the food AA_1.

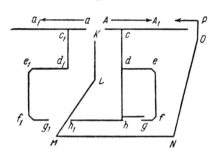

Fig. 64. Scheme of the experiment with
the screen with the complicated attachment. (Legend—see Fig. 62 and the text).

If the animal runs in the right direction in this experiment too,
the task is complicated still further: a complicated attachment
$c_1 d_1 e_1 f_1 g_1 h_1 hgf edc$ (Fig. 64) is joined to the screen. Not only is the
animal's path lengthened, but part of it (LM) is directed toward the
side opposite the movement of the food (AA_1). When the food moves
along the path aa_1, the screen dh is moved to the position d_1h_1, and
the screen hg into the position h_1g_1.

In order to go around the screen on the side toward which the
bowl of food has moved, the animal must obviously retain the track
of the acting stimulus, reflecting the change in its position in relation to the immobile environment.

The association formed (between the change in the position of
the stimulus and the environment in which the experiment is taking
place) is undoubtedly an essential condition for finding the stimulus.
In each successive experiment with the screens (the screen alone,
the screen with the simple attachment, and the screen with the
complicated attachment) the animal's nervous system must keep
track of the direction of movement of the food in relation to the
increasingly complicated change in the direction of the path of its
own movement.

Objective indices of the animal's behavior as it solved this
problem were: 1) the fact that the animal ran to the side toward
which the full or empty feeding bowl moved; 2) the time taken to
run to the feeding bowl (from the beginning of movement of the
bowl); 3) the latent period of the beginning of running (from the

moment the animal removes its head from the gap to its first step). The whole path taken by the animal was recorded.

The experiments with each animal were carried out twice a week. On the day of the experiment, the tests in which the food was moved behind the screen were performed 4–6 times. The experimental and control feeding bowls were moved on the day of the experiment an equal number of times toward both sides of the screen. At the beginning of the experimental period the movement of the bowls to the right and left sides was alternated. Subsequently, in order not to create a conditioned reflex to the alternation of the movement of the bowls to one side and then to the other, the stereotype was destroyed.

Experiments with pigeons. In the experiments in which food moved behind the screen, the reactions of the pigeons in relation to the moving food could be divided into three periods. In the first and very short period some birds attempted to seek the food which had disappeared from their field of vision. Their search movements were directed to the side toward which the food was moving. Being unable to reach the food as it moved away, four of the seven pigeons used in these tests, after removing their head from the gap, walked for a distance of 10–13 cm alongside the screen toward the side of the movement of the food, making as they did so characteristic movements with their whole body of "poking" through the screen. These search movements lasted a short time only (2–10 seconds). In no case did they enable the bird to walk around the screen on the side toward which the food was moving, and they were extinguished in the course of the first experiment, i.e., after 4–6 movements of the food behind the screen. As an example of the search extrapolation-reflex movements during the first experiment we may reproduce the notes of an experiment with pigeon No. 4.

Experiment with screen No. 1, on February 2. Velocity of movement of feeding bowls 3–4 cm/sec; length of path 30–40 cm.

16 hr 35 min. Pigeon pecked food from the feeding bowl through the gap. When the bowl moved away and the bird could no longer reach the food, it withdrew its head and made a movement with its whole trunk toward the direction of movement of the feeding bowl containing food for a distance of 10–12 cm along the screen, as if it were trying to poke its way through; after 2–3 seconds it went away from the screen and did not go around it.

16 hr 40 min. Pigeon pecked food from the feeding bowl. When the bird was unable to reach the food it withdrew its head and for the next 10 seconds made movements of "poking" its way through the screen, with its whole body, in the direction of movement of the food, after which it went away from the screen, described a loop, again approached the screen and made further "poking" movements through it on the side toward which the food had moved, and finally it went away from the screen without going around it.

In the remaining two cases of movement of the food in this experiment the pigeon also made clear searching "poking" movements on the side toward which the food had moved, but in no case did the bird go around the screen.

The second period in the behavior of the pigeons was characterized by absence of any deliberate searching for food. As soon as the bowl of food was moved away, the pigeons removed their head from the gap and began to walk in different directions. Sooner or later, going around the screen by chance, they reached the bowl containing the food. After the birds had walked around the screen several times in this manner, they developed a conditioned reflex walk around the screen on one side and an approach to the food on the side where it was accidentally found. By way of example we may produce the notes of an experiment with pigeon No. 4.

Experiment No. 3, on March 9. Feeding bowls with food moved at a velocity of 2.0–3.5 cm/sec; length of path 35–50 cm.

9)* 12 hr 50 min. Feeding bowl with food moved to the right. As soon as the pigeon found that it could not reach the food it removed its head from the gap, walked in front of the screen, passed behind it, and encountered the feeding bowl (containing food).

10) 13 hr. Feeding bowl with food moved to the left. The bird went away from the gap, then went back to it, walked the full length of the screen on the right, and after 35 seconds reached the empty feeding bowl.

Here we can see the beginning of the formation of a conditioned reflex walk along the whole length of the screen on the right side.

The third period in the behavior of the pigeons was characterized by a clear and constant conditioned reflex walk along the whole length of the screens on one side, irrespective of the direction in which the food moved.

As an illustration we may cite the behavior of the same pigeon (No. 4) in the 20th experiment with that bird after the 61st movement of the feeding bowls behind the screen.

Experiment No. 20, on June 21. Velocity of movement of feeding bowls 2–6 cm/sec; length of path 20–60 cm.

62) 14 hr 13 min. Feeding bowl with food moved to the left. As soon as the pigeon found that it could not reach the feeding bowl, it removed its head from the gap, moved back a distance of 30–40 cm from the screen, and ran to the side in the direction of movement of the empty feeding bowl, ran the whole length of the screen on the right, and reached the empty feeding bowl, in 10 seconds.

63) 14 hr 15 min. Feeding bowl with food moved to the right. The pigeon removed its head from the gap and after 1.5 seconds ran to the side toward which the bowl of food was moving, went around the screen and, in 15 seconds, reached the food.

64) 14 hr 18 min. Feeding bowl with food moved to the left. The pigeon removed its head from the gap and immediately went away to the side opposite to the direction of movement of the food, passed around the screen and, in 12 seconds, reached the empty feeding bowl.

65) 14 hr 20 min. Feeding bowl with food moved to the right. The pigeon removed its head from the gap and immediately went away, described a loop, and walked to the side toward which the bowl of food was moving, passed around the screen on the right side and, in 17 seconds, reached the bowl of food.

---

*The figure standing in front of the number indicating the time denotes the serial number of the movement of the feeding bowl from the beginning of the experiments with the particular animal.

In all seven pigeons taking part in the experiment a conditioned reflex was created, in the form of walking around the screen on one side. The conditioned reflexes formed were very stable and were extinguished with great difficulty by means of the constant movement of the feeding bowl to the side opposite to the direction of the conditioned reflex walk. In pigeon No. 7, for example (a street pigeon which we reared in the laboratory), the conditioned-reflex walk around the screen on the right side was not extinguished even after 132 movements of the feeding bowl containing food toward the left side of the screen only.

If the conditioned-reflex walk around one side of the screen could be extinguished, then the pigeon generally stopped walking around the screen. An example of this was pigeon No. 5, in which a previously formed conditioned-reflex walk around the screen on the right side was totally extinguished.

Pigeon No. 5. Experiment No. 10 on September 24, with extinction of the conditioned-reflex walk around the screen on the right side. Velocity of movement of feeding bowl 1 cm/sec. Length of path 10 cm.

36th extinction. 15 hr 20 min. Feeding bowl with food moved to the left. The pigeon removed its head from the gap and stood there for 18 seconds, after which it began to flap its wings and to move away from the screen.

37th extinction. 15 hr 22 min. Feeding bowl with food moved to the left. The pigeon removed its head from the gap, stood there for 8 seconds, and then walked to the side toward which the empty bowl was moving, but turned away from the screen and did not approach it again.

The following conclusion may be drawn from these experiments with pigeons. At the very beginning of the experiments in which the food was moved behind the screen, in some birds obvious searching movements toward the side of movement of the food could be detected. This attempt to find the food which had disappeared from the pigeon's sight was extinguished in the bird after a few movements of the bowl.

In the simplest variant of the experiment with pigeons, elements of rapidly extinguished extrapolation reflexes could thus be discerned, but in no case did these lead to walking around the screen on the side toward which the food was moving. Gradually, as a result of the chance walking around the screen on one side or the other, and of the discovery of food there, a stable conditioned-reflex walk around the screen on one side was established. After extinction of the conditioned-reflex walk around one side of the screen, the pigeons generally ceased to walk around the screen on either side. It is evident that extrapolation reflexes play an extremely small part in their behavior; conditioned-reflex connections are of principal importance in their adaptation.

Experiments with ducks. In none of the five ducks used in our experiments did we find definite movements to the side of the feeding bowl containing food as it moved away from the gap, behind the screen. Immediately after the moving of the feeding bowls, the birds as a rule went away from the gap and in most cases stayed a short distance from the screen. Finally, however, conditioned-reflex walking around the screen on one side was established in each duck as a result of chance performance of such excursions by the birds, and the discovery of food behind the screen. These walks were extinguished as a result of the movement of the food to the side opposite to that on which the birds walked around the screen. When the ducks had walked three times in a row around the screen on the side toward which the food was moved, we again began to move the bowl of food to both sides. Under these circumstances one duck (Belyanka) went around the screen several times on the side toward which the bowl of food was moved. In the next experiment, however, this duck again began to walk around the screen on one side only. The remaining ducks either stopped walking around the screen altogether or they continued to walk around it on one side only.

These experiments showed that the ducks mainly used a conditioned-reflex walk around the screen on one side in order to reach the food; the extrapolation reflex was of no essential importance for the finding of the concealed stimulus.

Experiments with fowls. In the very first experiment the fowls tried to go around the screen. In this first experiment, for instance, in which 24 fowls were used (the food was moved behind the screen 4–8 times), the birds walked around the screen and up to the feeding bowl (with or without food) 125 times, twice they flew over the screen, four times they squeezed through the gap in the screen, and only in 24 cases did the birds not reach the feeding bowls.

Another characteristic feature of the behavior of the fowl, which was clearly revealed in this experiment, was that the first searching movements after the bowl of food had been moved away behind the screen and the bird had taken its head from the gap were directed in most cases to the side toward which the bowl of food was moved. For instance, movements of the bird to the side toward which the feeding bowl containing food was moved were observed in 106 cases, and to the side of movement of the empty bowl in 38 cases, while in 11 cases the birds immediately went away from the screen.

Although the first movement was more often directed toward the side of movement of the food, this did not lead to a significant increase in the frequency of walking around the screen on the side toward which the bowl of food was moved. Of the 125 cases of walking around the screen in the first experiment, in 66 cases this was on the side of movement of the bowl of food, and in 59 cases—on the side of the empty bowl. An example is described below.

Experiment with the cock Spornyi. March 5. Experiment No. 1. Velocity of movement of feeding bowls 7–8 cm/sec. Length of path 35–40 cm.

1) 17 hr 58 min. Feeding bowl with food moved to the right. When the cock was no longer able to reach the bowl of food, it removed its head from the gap, stood there for 6 seconds, and walked for 25–30 cm to the side toward which the feeding bowl was moving, after which it returned to the gap and again walked to the side of the moving food bowl, went around the screen and, in 30 seconds, reached the bowl containing the food.

2) 18 hr 01 min. Feeding bowl with food moved to the left. The bird took its head from the gap and, after 2.5 seconds, walked for 20–25 cm to the side of movement of the bowl containing the food, making intensive movements with its head as it did so, as if trying to poke a way through the screen (Fig. 65), after which it returned to the gap and walked to the side of movement of the bowl containing food, went around the screen and, in 23 seconds, reached the bowl of food.

3) 18 hr 04 min. Feeding bowl with food moved to the right. The cock removed its head from the gap and, after 3 seconds, walked to the side of movement of the empty bowl, walked around the screen and, in 11 seconds, reached the empty bowl.

4) 18 hr 06 min. Feeding bowl with food moved to the left. The cock removed its head from the gap and, in 4 seconds, walked away for 30 cm to the side of movement of the bowl with food, then ran to the side of movement of the empty bowl, walked around the screen and, in 18 seconds, reached the empty bowl.

5) 18 hr 08 min. Feeding bowl with food moved to the right. After 1 second the bird went away from the gap and walked to the side of movement of the bowl with the food, making "poking" movements through the screen with its head, walked around the screen and, in 9 seconds, reached the bowl containing the food.

6) 18 hr 10 min. Feeding bowl with food moved to the left. The cock removed its head from the gap, and after 2 seconds walked to the side of movement of the empty feeding bowl, reached the end of the screen but did not go around it, turned back and walked to the side of movement of the bowl of food and, after going around the screen, reached the bowl containing the food.

In this experiment the cock tried to seek the food along the screen on the side toward which the bowl containing food was moved. These attempts were especially obvious during the second and fifth movements of the bowl, during which the bird not only walked to the side of movement of the food, but also tried to poke its way through the screen to the food which it could not see.

In the first day of the experiment, the fowls thus showed that they could develop an association between the food and the screen. This association was formed on the basis of the retention of the track of the stimulus, once it had acted, in their nervous system. However, no use was made of the direction of movement of the stimulus, which was detected by these birds, in their search for the food. Extrapolation reflexes were not sufficiently developed in the fowls to enable them to obtain the most adequate solution of the task presented to them.

Fig. 65. Cock searching for food near the gap in the screen.

After the second or third day of the experiment, the behavior of the fowls altered significantly. This was shown, first, by the fact that the searching movements in the direction of movement of the bowl containing the food were extinguished; second, by the fact that the majority of birds began to walk around the screen on any side, irrespective of the direction of movement of the food. By way of example we present the record of experiment No. 3 with the same cock, Spornyi.

March 12. Experiment No. 3 with the screen. Velocity of movement of feeding bowls 6–8 cm/sec. Length of path 30–40 cm.

13) 14 hr 31 min. Feeding bowl with food moved to the right. The cock at once walked toward the left side, went around the screen and, in 9 seconds, reached the empty feeding bowl.

14) 14 hr 33 min. Feeding bowl with food moved to the left. The bird removed its head from the gap after 3 seconds, walked toward the left side and, in 10 seconds, reached the feeding bowl containing food.

15) 14 hr 35 min. Feeding bowl with food moved to the right. The cock removed its head from the gap, after 1 second walked to the left side, and in 5 seconds reached the empty feeding bowl.

16) 14 hr 37 min. Feeding bowl with food moved to the left. The bird removed its head from the gap, after 1 second walked to the left side, and in 8 seconds reached the feeding bowl containing the food.

17) 14 hr 40 min. Feeding bowl with food moved to the right. The cock removed its head from the gap and, after 1.5 seconds, turned toward the side of movement of the feeding bowl with the food, but walked toward the opposite (left) side and, in 8 seconds, reached the empty feeding bowl.

18) 14 hr 42 min. Feeding bowl with food moved to the left. The cock removed its head from the gap, and after 1.5 seconds walked to the left side, went around the screen and, in 9 seconds, reached the bowl containing the food.

It may be seen from the record of this experiment that the behavior of the cock had become very uniform by the time of the third experiment. It no longer attempted to seek food in the direction of its movement, and only conditioned-reflex walking around the screen on the left side was observed, irrespective of the direction of movement of the food.

Of the 24 fowls taking part in our experiments, only five failed to develop a conditioned reflex walk around the screen on one side. In the remaining 19 birds, after the third to the fifth experiments stereotyped walking around the screen on one side was established, irrespective of the direction in which the food was moved. In some birds, however, in spite of the obvious conditioned-reflex walking around the screen on one side, the extrapolation movements toward the side of movement of the food were preserved. Each time the food was moved to the side opposite to that on which a conditioned-reflex walk around the screen had been developed by the bird, it nevertheless made a few steps toward the side of movement of the food, although it subsequently walked around the screen on the opposite side.

Experiments in which the conditioned-reflex walk around the screen on one particular side was extinguished showed differences in the behavior of the fowls. In some individuals, after extinction of the conditioned-reflex walk around the screen on one side, this was replaced by a conditioned-reflex walk around the screen on the other side (on the side toward which the food moved during extinction); in others we observed the appearance of extrapolation-reflex movements toward the side of the movement of the food and walking around the screen on the side of movement of the food, i.e., the most adequate solution of the problem assigned to the birds under the experimental conditions. After extinction, some fowls gave up walking around the screen.

As an example of the replacement of the conditioned-reflex walking around the screen on one side by an analogous walking around the other side, we may mention the experiments with the hen Sova. In this hen a conditioned-reflex walk around the screen on one side (the left) had been formed, irrespective of the side to which the food was moved.

This unilateral conditioned-reflex walking around the screen was extinguished by moving the food only to the right side of the screen. Complete extinction of the conditioned-reflex walking around the screen on the left side and the beginning of a conditioned-reflex walking around the screen on the opposite side (as indicated by three walks around the screen on the opposite side) were attained in the 11th experiment with extinction. Subsequently the feeding bowl containing the food began to be moved in both directions. The hen now began to walk around the screen only on the right side. This conditioned-reflex walking around the screen was very stable and appeared persistently in spite of attempts at its extinction.

An example of the appearance of correct walking around the screen after extinction of the conditioned-reflex walk around one side is given by the experiments with the cock Spornyi. The extrapolation movements and walking around the screen on both sides observed in the first experiment with this bird were replaced in the third experiment by conditioned-reflex walking around the screen only, and this on the left side. After the cock had walked around the screen on the left side 18 consecutive times (by our adopted standard), we began to carry out extinction of the conditioned-reflex walking around the screen which had been formed. By the time of the third experiment with extinction, the bird walked around the screen three times in a row on the right side.

Subsequently we began to move the feeding bowl with the food toward both sides. The cock began to make movements toward the side of the bowl containing the food, which in some cases terminated in correct walking around the screen on the side toward which the food was moved.

After seven experiments the cock, despite the change in the usual stereotype of the experiment (alternation of movement of the bowl of food first to one and then to the other side), began to walk around the screen every time on the side toward which the bowl containing the food was moved.

As we have pointed out above, a conditioned-reflex walk around the screen on one particular side was not established in all the fowls taking part in the experiments. In five birds no conditioned reflex was formed; in the course of all the subsequent work with these birds they preserved their search movements in the direction of movement of the food, which were gradually consolidated and led to the correct walking around the screen on the side toward which the food was moved. As an example we may cite the experiment with hen No. 1724, which by the time of the seventh experiment had begun to walk around the screen in every case on the side toward which the feeding bowl with the food was moved.

February 2. Experiment No. 7 with the screen. Velocity of movement of feeding bowls 7–9 cm/sec. Length of path 35–40 cm.

49) 16 hr. Feeding bowl with food moved to the left. The hen walked to the left side, walked around the screen and, in 12 seconds, reached the bowl of food.

50) 16 hr 02 min. Feeding bowl with food moved to the right. The bird removed its head from the gap and walked to the right side, walked around the screen and, in 10 seconds, reached the bowl containing the food.

51) 16 hr 04 min. Feeding bowl with food moved to the left. The hen removed its head from the gap and walked to the left side; in 10 seconds it reached the bowl containing food.

52) 16 hr 06 min. Feeding bowl with food moved to the right. The bird removed its head from the gap, walked to the right side and, in 10 seconds, reached the feeding bowl containing food.

53) 16 hr 08 min. Feeding bowl with food moved to the left. The bird walked to the left side of the screen and around it and, in 10 seconds, reached the feeding bowl with the food.

54) 16 hr 10 min. Feeding bowl with food moved to the right. The bird went to the right side of the screen and around it and, in 11 seconds, reached the feeding bowl with the food.

55) 16 hr 12 min. Feeding bowl with food moved to the left. The hen walked to the left side of the screen and around it and, in 11 seconds, reached the feeding bowl with the food.

56) 16 hr 14 min. Feeding bowl with food moved to the right. The bird walked to the right side of the screen and around it and, in 11 seconds, reached the bowl containing the food.

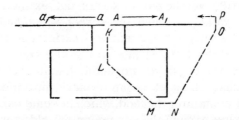

Fig. 66. Scheme of the experiment with the complicated attachment with two openings (KLMNOP —movement of the animal by the shortest route to the food).

The change in the stereotype affecting the sequence of movement of the feeding bowl to alternate sides of the screen did not disturb the correct walking by the birds around the screen on the side toward which the bowl of food was moved.

The experiment was then made more complicated. An attachment was fitted to the screen (Fig. 63). In the first experiment the hen still made three correct walks out of four. In the subsequent experiments, when it had become obvious that the hen walked around the screen far more often on the side of movement of the bowl containing food, a new complication was introduced: a complex attachment was added to the screen with the simple attachment (Fig. 64). In the first experiment, in all four cases the hen now walked around the screen on the side toward which the bowl containing food was moved.

Further experiments showed that, in the overwhelming majority of cases, the hen walked around the screen with the complex attachment on the side toward which the bowl containing food was moved.

Similar results were also obtained with the other four birds, which also were able to walk around the screen with the complex attachment on the side of movement of the bowl of food. In order to do this, however, many weeks of individual experience were required. On the basis of the initial extrapolation reflexes, as a result of prolonged individual experience, conditioned reflexes of walking around the complex screen on the side of movement of the bowl of food were formed. The fact that these walking movements were automatized as conditioned reflexes was clearly shown when we changed from experiments with the very complex screen to a simpler variant of the experiment.

In this variant of the experiment, we introduced a partition into the complex attachment, and made two openings which made it possible for the birds to reach the food by a shorter route (Fig. 66) than in the main variant of the experiment (Fig. 64). The changeover to the modified task, however, at first was quite beyond the power of the hens and they made many errors.

As an example we may cite an experiment with the cock Levyi, which walked almost faultlessly around the screen with the complex attachment on the side toward which the bowl containing the food was moved. In experiment 18, on October 22, for instance, 14 out of 15 times this bird walked around the screen on the side of movement of the bowl containing the food. In the next experiment, on October 25, both doors in the new complication were opened.

Experiment on October 25 (both doors open). Velocity of movement of the feeding bowls 4–10 cm/sec. Length of path 40–50 cm.

14 hr 18 min. Feeding bowl with food moved to the left; after 2.5 seconds the cock turned to the side of movement of the food, passed through the door closest to the food, but went in the direction of movement of the empty feeding bowl and, in 18 seconds, reached the empty feeding bowl.

14 hr 20 min. Feeding bowl with food moved to the right. After 2 seconds the bird turned to the side of movement of the food, passed through the door closest to the food and, in 15 seconds, reached the feeding bowl with the food.

14 hr 23 min. Feeding bowl with food moved to the left. After 2 seconds the cock turned to the side of movement of the food, passed through the door closest to the food, but went to the side of movement of the empty feeding bowl, which it reached in 15 seconds.

16 hr 26 min. Feeding bowl with food moved to the right. The cock at once turned to the side of movement of the food, passed through the door closest to the food and, in 17 seconds, reached the food.

14 hr 30 min. Feeding bowl with food moved to the left. After 2 seconds the bird turned to the side of movement of the food, but passed through the door furthest from the food, walked to the side of movement of the empty feeding bowl, and reached this bowl, in 16 seconds.

In this experiment the cock made three wrong walks around the screen in five cases of movement of the food. In the course of three subsequent experiments, the cock solved this problem and began to walk to the other side of the screen by the shortest route. When we again changed to the original, more difficult variant of the experiment (Fig. 64), the bird began to walk around the screen incorrectly.

These experiments just described show that the correct walking movements in the direction of movement of the food, brought about by means of extrapolation reflexes, were subsequently consolidated as a result of conditioned-reflex automatization. As a result of this consolidation, when the stereotyped experimental conditions were changed, difficulties were experienced in changing to a new method of solution of the assigned problem.

Fig. 67. Crow running in the direction of movement of the food behind the screen.

These experiments with fowls thus revealed the extraordinary variation in the ability of these birds to find the food moving behind the screen. The search for the food in the direction of its movement behind the screen, which was observed in the course of the first experiment with the fowls, was replaced in the majority of the birds by a conditioned-reflex stereotyped walking around one particular side.

In all three variants of the experiment with the screen, of increasing complexity, correct walking movements around the screen were possible in those fowls in which the extrapolation reflexes were sufficiently well defined and no conditioned-reflex walking movement around one particular side was formed.

On the basis of these regularly performed walks around the screen on the side of movement of the food, a pattern of behavior is formed which, as a result of the "overgrowth" of the extrapolation reflexes by conditioned-reflex components, is gradually automatized. As a result of the conditioned-reflex automatization of the walking movements around the screen, however, this walking is carried out not on any side, but on the side of movement of the feeding bowl containing the food, which enables the bird to obtain food in every case of movement of the bowl containing food. This method of reaching the food is more in keeping with the experimental conditions.

Experiments with birds of the crow family. In the first experiment in which the food was moved behind the screen, the behavior of these birds differed considerably from that of the pigeons, ducks, and fowls. After the first experiment with movement of the feeding bowl behind the screen, the birds of the crow family, in the overwhelming majority of cases, walked around the screen on the side of movement of the feeding bowl containing the food (Fig. 67).* In the first experiment with 14 birds of the crow family, of 70 movements of the feeding bowls, the birds went around the screen on the side of movement of the feeding bowl with the food in 60 cases (85.7%). After the second experiment, hardly any mistakes were made in going around the screen. Only one crow, Varya, began to walk around the screen on one side only. As an example of correct walking around the screen, we present the record of the first experiment with the magpie Bezymyanka.

April 26. Experiment No. 1 with the screen. Velocity of movement of the feeding bowls 3–4 cm/sec. Length of path 30–40 cm.

13 hr 30 min. Feeding bowl with food moved to the left. As soon as the magpie found that it could no longer reach the feeding bowl with the food, it removed its head from the gap and paced up and down for 3 seconds near the gap, after which it ran to the left side of the screen, ran around it and, in 14 seconds, reached the feeding bowl with the food.

13 hr 45 min. Feeding bowl with food moved to the right. As soon as the bird found that it could no longer reach the feeding bowl with the food, it ran to the side of movement of the food for 20–30 cm, returned to the gap, peeped through, and again ran to the side of movement of the feeding bowl with the food, ran around the screen on the right side and, in 17 seconds, reached the feeding bowl with the food.

13 hr 50 min. Feeding bowl with food moved to the left. As soon as the magpie found that it could no longer reach the food, it removed its head from the gap, paced up and down by the gap for 3 seconds, peeped through it, and at once ran to the side of movement of the food, ran around the screen on the left side and, in 15 seconds, reached the feeding bowl containing the food.

13 hr 55 min. Feeding bowl with food moved to the right. The bird removed its head from the gap and immediately ran to the side of movement of the feeding bowl with the food, ran around the screen on the right side and, in 11 seconds, reached the bowl with the food.

*In two other cases the crow flew over the screen on the side of movement of the feeding bowl containing the food.

In the solution of the task next in degree of complexity (the screen with the attachment), some of the experimental birds had considerable difficulty. Of 56 movements of the feeding bowls with the food behind the screen with the attachment in the first experiment there were 37 cases (66.6%) of walking around the screen on the side of movement of the feeding bowl with the food. In subsequent experiments (third or fourth), in most birds the number of correct walking movements around the screen with the attachment on the side of movement of the bowl containing the food increased. In some birds, however, the opposite was observed: difficulty in the solution of the task was greater in the subsequent experiments. This was shown not only by the relatively large number of incorrect walking movements, but also by the tendency to walk around the screen with the attachment on one particular side.

The magpie Bezymyanka, for example, which walked correctly around the screen with the attachment in all cases when the bowl with the food was moved in the first experiment, in subsequent experiments began to walk around it only on the right side.

Difficulty in walking around the screen with the attachment on the side of movement of the bowl containing the food was experienced by one of our crows, Pava (a very tame crow with very high food excitability), for during the solution of this problem it developed signs of fear of the bowl containing the food.

May 24. Experiment No. 2 with the screen and attachment (the crow Pava). Velocity of movement of feeding bowls 6–10 cm/sec. Length of path 30 cm.

5) 16 hr 30 min. Feeding bowl with food moved to the left. Pava immediately ran out of the attachment, ran to the left side of the screen and around it, but then stopped and, with obvious signs of "fright," looked at the feeding bowl, ran back to the attachment, again ran around the screen on the side of movement of the feeding bowl with the food, again showed signs of defensive behavior, returned to the edge of the screen, looked behind it, looked at the food and then ran away from the screen.

6) 16 hr 40 min. Feeding bowl with food moved to the right. The crow at once jumped from the attachment, ran to the right side of the screen and reached its end, but, showing signs of passive defensive behavior, turned back, ran to the opposite side of the screen, again returned to the right side of the screen, and slowly, with great "caution," reached the feeding bowl with the food (in 48 seconds).

7) 16 hr 44 min. Feeding bowl with food moved to the left. The bird at once ran from the attachment, ran to the left side of the screen, ran around it and, in 14 seconds, reached the feeding bowl with the food.

8) 16 hr 47 min. Feeding bowl with food moved to the right. Pava immediately jumped from the attachment, ran to the right side of the screen and around it and, in 8 seconds, reached the feeding bowl with the food.

The next most difficult task (coping with the screen with the complex attachment) was even harder for the birds to solve. In the first experiment, in which seven birds of the crow family were used, there were 14 (out of 27) walking movements around the screen on the side toward which the bowl containing the food was moved, ten such movements on the side of the empty bowl,

Table 45. Experiments with the Screen with the Complex Attachment and with Birds of the Crow Family

Name	Species	Result of experiments with the screen with the complex attachment
Pava	Crow	in the 2nd experiment, in all 4 cases walked around the screen on the side of movement of the food; in subsequent experiments marked fear of the experimental conditions developed
Varyag	Crow	in the 2nd experiment, in all cases walked around the screen on the side of movement of the food; in subsequent experiments walked correctly around the screen on the side of the direction of movement of the feeding bowl with the food
Varya	Crow	in all experiments walked around the screen on one side only; after extinction—walked correctly around on the side of movement of the feeding bowl with the food
Kralya	Crow	in the 2nd experiment, in 5 (of 6) cases walked around on the side of movement of the feeding bowl with the food; in subsequent experiments walked around the screen mainly on the same side
Zhulya	Magpie	in all experiments walked around the screen on one side only
Artemka	Magpie	in the 6th experiment, in 5 (of 6) cases walked around on the side of movement of the food; in subsequent experiments walked around correctly
Tsyganka	Rook	in the 3rd experiment, in 5 (of 6) cases walked around on the side of movement of the food; then began to walk around the screen on one side only

and in three cases the birds did not reach any of the bowls behind the screen. In the subsequent experiments the result varied in the different birds (Table 45).

It can be seen from Table 45 that five birds walked correctly around the complex screen after the 2nd–6th experiments (Pava, Varyag, Kralya, Artemka, and Tsyganka). Two birds (Varya and Zhulya) from the very beginning walked around the complex screen on one side only.

Of the five birds which began by walking around the complex screen on the side toward which the feeding bowl containing food was moved, in three these correct walking movements around the screen were quickly deranged. The rook Tsyganka and the crow Kralya began to walk around the screen mainly on one side, and the crow Pava, after solving the problem correctly, developed a severe phobia* in relation to the experimental conditions (fear of the screen and feeding bowls).

Those of our birds which after a little individual experience were able to walk around the complex screen on the side toward which the feeding bowl with the food moved did so as a result of

*Phobia—an obsessive, neurotic, passive defensive reaction.

"overgrowth" of extrapolation reflexes (which were evidently inadequate to provide the correct solution of the problem—walking
around the screen correctly) by conditioned-reflex components.

As an example we may cite the experiments with the magpie
Artemka. This bird in its first experiment (March 1, 1957) in every
case walked around the screen alone and around the screen with
the attachment on the side toward which the bowl with the food
moved. When the complex attachment was introduced into this
experiment, however, the magpie was observed to have difficulty
in walking around the screen on the side toward which the food
moved.

In the next experiment (March 5) the bird in every case walked
around the complex screen on the right side only. When the feeding
bowl moved to the left, however, the magpie ran from the complex
attachment in the direction of movement of the bowl with the food
but did not run around the screen; it turned back and ran around it
on the side toward which the empty feeding bowl moved.

In the two subsequent experiments on March 8 and 12, changes
took place in the behavior of the magpie. It walked around the screen
on the right side faultlessly, but in 2 cases (of 6) it also correctly
ran around the screen on the left. Its individual experience was
adequate, and beginning with the following experiments (March 15
and 19), the bird began to walk around the complex screen correctly
on the side of movement of the bowl containing food, whatever side
it happened to be. This example shows that the formation of a
complex act of behavior was taking place in the bird. Extrapolation
reflexes alone were inadequate for walking around the screen
with the complex attachment; prolonged individual experience was
also necessary.

Walking around the complex screen on the right side was along
a route which the magpie already knew well enough, but walking on
the left side was not yet sufficiently well reinforced by individual
experience. Every time the food moved to the left, therefore, as a
result of the adequate extrapolation reflex the bird ran to the left,
but did not reach the food. It is obvious that the track of the direction of movement of the food on this side was not retained in its
nervous system long enough for the bird to run the whole length of
the path. This necessitated in addition a conditioned-reflex "knowledge" of the path leading to the food. Walking around the screen on
the left side on two occasions (March 8 and 12) was sufficient to
enable the bird in subsequent experiments to walk around the complex screen on the side toward which the food was moved. However,

if the magpie's extrapolation reflexes were not sufficiently established, walking around the complex screen on the right side was
undoubtedly given conditioned-reflex reinforcement, and the bird
walked around the complex screen on the right side only.

In animals with relatively feebly developed extrapolation reflexes, the conditioned reflexes formed in response to the feeding
bowl on a particular side of the screen lead to stereotyped walking
around the screen on one side only, but relatively well-developed
extrapolation reflexes (as in birds of the crow family) evidently
become supplemented by conditioned reflexes, enabling the birds
to walk around the screen on the side toward which the food is
moved. Although in birds of the crow family conditioned reflexes
undoubtedly play a role in their walking around the screen on the
side of movement of the food, these birds, in contrast to the fowls,
are able rapidly to modify their behavior to suit the changes in
this task.

When we changed to experiments in which both doors in the
complex attachment were open, the birds of the crow family immediately ran through the door closest to the food and then ran
around the screen on the side toward which the feeding bowl with
the food moved. By way of example we describe an experiment with
the magpie Artemka.

Experiment on March 29. Velocity of movement of feeding bowls behind the screen
7–10 cm/sec.

39) 11 hr 54 min. Feeding bowl with food moved to the left. The bird at once jumped
back from the gap in the screen, then peered into it, ran through the complex attachment,
and reached the feeding bowl with the food in 14 seconds.

40) 11 hr 56 min. Feeding bowl with food moved to the right; the magpie turned to the
side toward which the food was moving, ran through the complex attachment, and reached
the feeding bowl with the food in 7 seconds.

At 11 hr 58 min and 12 hr 03 min, two similar experiments were carried out, in which
the bird ran around the screen on the side toward which the feeding bowl with the food
was moved. Both doors in the complex attachment were then opened.

43) 12 hr 07 min. Feeding bowl with food moved to the left. The magpie at once ran
out through the door nearest to the food, and reached the feeding bowl with the food in
8 seconds.

44) 12 hr 10 min. Feeding bowl with food moved to the right. The bird at once turned
to the side toward which the food was moved, ran out of the door nearest to the food, and
reached the feeding bowl with the food in 7 seconds.

It can be seen clearly from this experiment that as soon as the
experimental conditions were altered, the magpie immediately
changed to the method of solution of the problem most suited to the
particular conditions, and ran out of the door nearest to the food.
Under these circumstances the birds of the crow family showed their
difference from the fowls, which could solve the simpler problem
only with difficulty, indicating that their solution of the more difficult
problem was automatized by a conditioned-reflex mechanism.

We have previously pointed out that the crow Pava developed manifestations of fear of the experimental situation during the experiments with the screen with the simple attachment. These manifestations first appeared, and then disappeared, but when we changed to the experiments with the complex attachment, the phobia of the experimental situation became much more severe, and it was impossible to work with the bird.

It is interesting to observe that in the first experiment with the complex attachment, Pava developed an obvious fear only of the left side of the screen. In the second experiment, in order to increase the food excitability of the bird, we added cottage cheese to the customary meat stimulus. The bird ate this mixture with extraordinary avidity, and in none of the four cases did it evidence fear as it went around the screen on the correct side and ran up to the feeding bowl.

This strain on the extrapolation-reflex activity was, however, too much for the bird's nervous system, and in the subsequent experiments (July 2 and 5) it developed a well marked phobia of the whole experimental situation. The bird began to be afraid of the right side of the screen as well as the left; it approached with trepidation food situated twenty or more centimeters from the screen. We give below an extract from the records of this experiment.

July 5. 12 hr 10 min. Pava would not venture into either the complex or the simple attachment, and was afraid of the whole experimental situation.

12 hr 53 min. The simple and complex attachments were removed. The bird was less afraid of the screen alone than with the attachments.

12 hr 58 min. Pava pecked at the food through the gap with extraordinary caution. After pecking, it immediately hopped away. The feeding bowl with the food was moved to the right. Velocity of movement of the food 7 cm/sec, length of path 35 cm. When the bowl of food began to move the crow hopped away from the gap in the screen, but although it ran to the side toward which the food had moved and went around the screen on the right side, it did not reach the food, but looked at it for 15 seconds and then went away.

13 hr 04 min. Feeding bowl with food placed at the outer side of the screen, near the gap. Pava went up to the bowl and pecked the food with no sign of fear.

13 hr 05 min. Feeding bowl placed behind the screen. Velocity of movement of feeding bowls 7 cm/sec; length of path 35 cm. Feeding bowl with food moved to the left. When it began to move, the magpie hopped away from the gap and ran to the side toward which the bowl of food moved; it went around the screen on the left side but did not reach the food; it ran to the opposite side of the screen, ran around it, reached the empty feeding bowl, and peered into it.

It is clear from these experiments that as a result of the solving of a task so difficult for the bird's nervous system (walking around the screen with the complex attachment on the side of movement of the food), Pava's normal nervous activity was disturbed. Fear of the moving food and of the experimental situation developed. This phobia developed particularly intensively in relation to the circumstances of the most difficult experiment (the screen with the complex attachment). It must be pointed out that Pava was a very tame crow, taken from a school biology group where it had been for several years. Pava lived in our laboratory for more than one year. The bird often was set free from its cage into the laboratory, and was never known to display the slightest sign of fear of anything in the laboratory.

After extreme strain of its extrapolation-reflex activity, the bird developed pathological behavior in the form of a definite phobia. Experiments on Pava were discontinued and it was decided to give the bird a long rest. Three weeks later (on July 26) Pava died.

We have made a detailed examination of the development of the pathological condition in this bird, because the appearance of a phobia during the solution of tasks presenting difficulty to the nervous system was observed in two more crows, Varya and Varyag.

Varya was the only crow which went around the screen (without attachments) on the right side only, irrespective of the direction of the movement of the food. Extinction of the conditioned reflex walking around the screen on this side led to the bird's walking correctly around the screen. In the next experiment, however, Varya began to be afraid to approach the feeding bowl with the food. The phobia was especially severe after the task had been made more difficult, and the crow had to pass around the screen with the attachment. The developing phobia could be treated successfully by means of sodium bromide, and the fear subsided.

Analysis of the causes of the development of a phobia in Varyag is complicated by the fact that during the experiment on January 21 (when it ate through the gap) the bird was frightened by an accidental sound stimulus. In the experiment on January 24 fear of approaching the food was exhibited only during its first movement behind the screen with the complex attachment; in the other five cases no traces of phobia were observed during movement of the food. In the experiment on January 31 fear of going around the screen appeared during the first two movements of the food behind the screen. In the next experiment (February 4) there were no signs of a phobia, but during five (of 6) movements of the feeding bowls behind the screen with the complex attachment, the crow passed around the screen on one side only (the right). Subsequently the experiments in which a phobia in relation to the situation periodically appeared and disappeared were replaced by experiments in which the crow passed around the screen with the complex attachment mainly on one side (the right). In the middle of March the crow began to pass around the screen only on the side of movement of the feeding bowl with the food; the phobia in relation to the experimental situation disappeared. It is very probable that the reaction of fear, caused by the accidental sound stimulus during the experiment on January 21, was maintained by the bird's neurotic state, which had developed as a result of the overstraining of its nervous system during the solution of the difficult problem. The fact that this problem was difficult for the bird's nervous system is shown by the constant tendency to walk around the screen on one side only—the conditioned-reflex automatization of the solution of the problem.

The physiological analysis carried out on dogs by Petrova (1935) in Pavlov's laboratory showed the important role of overstrain of the nervous system in the development of phobias, resulting from some previous nervous trauma induced by frightening stimuli. It is obvious that in our case the overstraining of the nervous system, caused by the necessity of solving the difficult problem, led to the prolonged preservation of the sequelae of the accidental fright.

The experiments with birds of the crow family may be summed up as follows. In the overwhelming majority of cases, even in the first experiment, these birds passed around the screen (without attachments) on the side of movement of the feeding bowl with the food. The subsequent complication of the experiment made it more difficult for the bird to walk around the screen on the side of movement of the food. The greatest difficulty arose in the experiments with the screen with the complex attachment. Although many birds began to walk around the screen on the side toward which the bowl with the food was moved by the time of the 2nd–6th experiments, the overstraining of their nervous system as a result of the necessity of using extrapolation reflexes in solving these difficult problems led to the development of signs of pathological behavior, to the development of a phobia. The difficulty in walking correctly around the screen with the complex attachment was also expressed by the fact that some birds began to walk around it on one side only, irrespective of the direction of movement of the bowl containing the food.

Experiments with rabbits. The behavior of our three rabbits in the first experiment in which they took part was fairly uniform. In the first and second experiments, during 14 (of 16) movements of the feeding bowls behind the screen, the rabbits removed their heads from the gap and ran in the direction of the bowl with the food. These

searching movements weakened, however, and after running a distance of 30–60 cm along the screen, the animals either simply ran
away from the screen or, even if they went around it, ceased looking
for the food and did not go up to the feeding bowl. In five of 16 cases
of movement of the feeding bowls, they went around the screen on
the side of movement of the feeding bowl with the food, but not
once did they reach the bowl; on one occasion the animal went
around the screen on the side opposite to the movement of the food;
in ten cases the rabbits did not go around the screen at all.

Further investigations showed individual differences in the
behavior of the animals studied.

The rabbit Kol'ka. In this rabbit we observed in the first experiments obvious extrapolation movements in the direction that the
food was moved. During the first six experiments the rabbit, in 19
of 25 cases, ran in the direction of movement of the food (although
it ran around the screen and reached the food in only seven cases);
it ran to the side opposite to the direction of movement of the food
in four cases, and in two cases it ran away from the screen at an
angle of approximately 45–50° in the opposite direction to the movement of the food.

After these six experiments with the rabbit no further experiments were carried out with it for 11 months. After this interval,
the animal again ran more often along and around the screen on
the side toward which the feeding bowl with the food was moved.
Beginning with the sixth experiment (after the interval), however,
this rabbit started to make conditioned-reflex walking movements
around the screen on one side only, irrespective of the direction
of movement of the food. By the time of the tenth experiment these
movements had become perfectly uniform in character: irrespective
of the direction of movement of the bowl containing the food, the
rabbit ran to the right side of the screen and up to the feeding bowl
(empty or containing food). We then began to extinguish its conditioned-reflex walking around the screen on the right side, i.e.,
we began to move the bowl with the food to the left side of the
screen. After 47 such movements of the feeding bowl the conditioned-
reflex walking of the animal around the screen on the right side
was extinguished: the rabbit in the course of one experiment walked
around the screen on the left side three times in succession and
reached the food. We then began to move the food alternately, according to our usual method, first to one and then to the other
side of the screen. Now, however, the rabbit began to pass around

the screen on the left side only, i.e., as a result of the extinction of the conditioned-reflex walking around the screen on the right side, it had been replaced by going around the screen on the left. Extinction of the conditioned-reflex walking around the screen was again begun, but this time on the left side. After 26 movements of the feeding bowl to the right side of the screen the conditioned-reflex walking around the screen on the left side was extinguished: three times in succession the rabbit went around the screen and up to the food on the right side. Now, however, when the food began to be moved to both sides, it was found that the rabbit persistently began to go around the screen only on the right side. At this stage the experiments with the rabbit Kol'ka were discontinued. It had become clear that the conditioned-reflex walking around the screen on one side only had been replaced after its extinction by a conditioned-reflex walking around the screen on the other side. The extrapolation reflexes, which were clearly observed in the first experiments with these rabbits, were completely inhibited by the conditioned-reflex walking around the screen on one side only.

The rabbit Serka. In this rabbit the extrapolation movements in the direction of movement of the food, which were observed in the first experiment, began to be replaced by the third experiment by a conditioned-reflex walking around the screen on only one side (the right). By the time of the 11th experiment they had become quite uniform in character: the rabbit ran around the screen on the right side every time, irrespective of the direction of movement of the food, and ran up to the feeding bowl (empty or containing food). We then began to extinguish the conditioned-reflex walking around the screen on the right side, i.e., to move the food to the left side only. This extinction took place slowly. Only after 115 movements of the food to the left did the rabbit pass around the screen twice in succession on the left side. We then began to move the feeding bowl with the food alternately to either side of the screen. The animal began to go around the screen correctly on the side of movement of the food. On the day after we discontinued the extinction, the rabbit each time went around the screen on the side to which the food was moved. When we were satisfied that, after extinction of the conditioned-reflex walking around the screen on one side only, an extrapolation-reflex method of going around the screen and reaching the food had been developed, the animal was presented with the task next in degree of difficulty, namely the screen with the simple attachment.

After the first experiment, in all cases of movement of the feeding bowls containing food, the rabbit began to walk around the screen with the attachment on the left side only. In the second experiment the attachment was taken away, but the animal continued to walk around the screen on the left side only. Consequently, as soon as the problem had become too difficult for the animal's nervous system, it changed over to a conditioned-reflex method of solution which, although less adequate (the animal obtained the food only every other time), was easier. The conditioned-reflex walking around the screen with the attachment, however, also affected the solution of the simpler problem, that of the screen without the attachment, in which previously the rabbit had passed around the screen on the side of movement of the bowl with the food; now it began to walk around on the left side only.

These unilateral walking movements around the screen without attachments and on the left side only did not, however, last long. After four movements of the feeding bowl to the right, the rabbit went in the direction of movement of the food. In the next two experiments it began to walk around the screen almost faultlessly, going each time in the direction of movement of the feeding bowl with the food. Such a method of solution of the problem, however, was nevertheless beyond the capacity of the nervous system of this rabbit. In the subsequent experiments the rabbit again began to walk around the screen on one side only (the direction of these walking movements now began to change from one experiment to another, and sometimes they were mainly on the left, sometimes mainly on the right side). Sometimes the rabbit did not run around the screen at all, but immediately after the movement of the bowls it ran away from the screen and did not come near it again. At the same time the animal developed obvious manifestations of phobia. When the rabbit ran up to the gap it did not begin at once to eat the food, but having thrust its head through the gap, it immediately ran away; it made a characteristic noise with its hindlimbs, which was evidence of the animal's excited state. It is significant that evidence of phobia and refusal to work were observed in those experiments in which the rabbit walked around the screen on the side toward which the bowl with the food was moved.

In subsequent experiments the tendency of the rabbit to pass around the screen mainly on one particular side was increased. Sometimes, for example, in the first half of the experiment it walked around on the right side, and in the other half of the experiment,

on the left side. Simultaneously with the development of a tendency
to walk around the screen on one side only, i.e., when the rabbit
changed over from an extrapolation-reflex method of solution of
the problem to a conditioned-reflex method, all the manifestations
of the phobia disappeared.

The rabbit Pyshka. In this rabbit, in the course of the first two
experiments we observed a tendency to run alongside the screen
in the direction of movement of the bowl containing the food. The
animal did not pass around the screen, however, and did not reach
the feeding bowl. It then began to walk around the screen mainly
on the left side. In the sixth experiment, during all six movements
of the feeding bowls the rabbit ran around the screen on the left
side. The conditioned-reflex stereotype of walking around the
screen on one side was nevertheless not found in this animal. In
the 10th and 11th experiments it began to walk around the screen
mainly on the side toward which the bowl containing the food was
moved. A more complicated task was then presented to the animal—
the screen with the simple attachment. This task was a strain on
the animal's nervous system. Random walking around the screen
with the attachment, on either side, with no semblance of order,
or a refusal to walk around the screen at all, and attempts to push
through the gap in the screen, were observed. It was decided to
test the action of sodium bromide against this background of obvious
difficulty in the solution of this task. The result of the experiments
was highly significant.

In the experiment on January 21, after the rabbit had walked
around the left side of the screen with the attachment five times in
succession, it was given 150 mg of sodium bromide. After an
interval of 45 minutes the experiment was resumed. It was found
that in five of six movements of the bowl with the food behind the
screen with the attachment the rabbit went around the screen in the
direction of movement of the food. Subsequently (January 24 and 28),
when the bowl of food was moved behind the screen with the attachment, in 10 of 14 cases the rabbit went around the screen on the
side toward which the bowl with the food was moved. A single administration of sodium bromide thus had a positive influence on the
solution of this difficult problem for the nervous system—to go
around the screen with the attachment on the side toward which
the food was moved.

The rabbit was unable to solve the next task in order of difficulty (to go around the screen with the complex attachment). From
the very first experiment the animal began to go around the screen

on only one side (the right), irrespective of the direction of movement of the food. After seven experiments in the course of which the rabbit walked around the screen mainly on the right side, the problem was simplified: the complex attachment was removed. It was found that the conditioned-reflex walking around the screen on the right side only was now preserved during the solution of the problem with the screen with the simple attachment, and even that of the problem with the screen without attachments.

After an interval of three weeks (March 12 to April 1), during which no experiments were undertaken with this rabbit, the animal again began to walk around the screen without attachments on the side toward which the bowl with the food was moved; it continued to walk around the screen with the attachment on the right side only.

However, the rabbit continued to walk around the screen without attachments on the side of movement of the bowl containing food for only a few experiments after the interval. The animal developed an obvious tendency to walk around the screen on the right side only. Meanwhile, in an experiment on May 6, the first signs of a phobia appeared—fear of the screen and the gap. After May 9 we began to move the feeding bowl only to the left side of the screen. Extinction of the conditioned-reflex automatism which had developed, and the necessity of solving the problem with the aid of extrapolation reflexes, were difficult for the rabbit. In the experiments on May 19 and 28 the rabbit exhibited a definite phobia in relation to the experimental situation. Fear was manifested principally to the left side of the screen, namely the side along which the food now moved. After the conditioned-reflex walking around the screen on the right side had been extinguished, however, and the rabbit had again begun to run around the screen on the side toward which the bowl with the food was moved, the phobia aroused by the experimental situation began to subside. The transfer from a conditioned-reflex method of solution of the problem to an extrapolation-reflex method was particularly difficult for the rabbit's nervous system, already exhausted by the previous experiments.

In the evaluation of the general features of the nervous activity of rabbits, as revealed by the experiments with the screen, the following remarks may be made. The extrapolation-reflex method of finding the food was expressed relatively weakly in these animals. Their attempts to find the food moving behind the screen by means of extrapolation reflexes, as manifested during the first experiments, showed a clear tendency toward replacement by conditioned-reflex walking around the screen. The strain on the nervous system

as a result of solving the problem by means of extrapolation reflexes led to the development of manifestations of a neurotic state.*

The experiments with the screen showed considerable differences in different species of animals in their method of seeking
food by moving away from a gap in a screen through which the
experimental animal had obtained information concerning the direction of movement of the stimulus. In some species of animals
(pigeons, ducks) this information was practically of no importance
in directing their search. The running movements in the direction
of movement of the food, observed in some pigeons during the first
experiments, were extinguished, and the birds developed conditioned-reflex walking movements around the screen on one side
only, irrespective of the direction of movement of the food. The
food, as it moved away from the gap, acted as a stimulus for the
animal to walk around the screen on one side only, and this conditioned reflex became stereotyped and automatized in character
in these animals, and was extinguished with difficulty.

In the case of fowls and rabbits, the information regarding the
direction of movement of the stimulus obtained through the gap was
now of some slight importance in the choice of direction of the
search for food. Among both fowls and rabbits, individuals were
encountered which made use of the information obtained to walk
around the screen on the side along which the stimulus was moved.
In order to be able to walk systematically around the screen on the
side of movement of the stimulus, however, a definite degree of
individual experience was necessary for them. Thus the food stimulus moving away from the gap may in the case of fowls and rabbits
act not only as a stimulus for the conditioned-reflex walking around
the screen on any one side, but may also give information to the
animal regarding the direction of its movement. It is evidently
easier for the nervous system of these animals to use the information thus obtained for the formation of a conditioned-reflex walk
around the screen on one side only, disregarding the direction of
movement of the stimulus.

*After this book was written, experiments with two more rabbits were concluded.
Both animals walked around the screen mainly on one side, irrespective of the direction
of movement of the food. After extinction of the unilateral walking around the screen, one
of the animals began to show correct walking around the screen on the side of movement
of the food. Signs of disturbance of the ordinary behavior appeared at the same time, as
shown by a general state of inhibition, and of "hanging about" near the gap after the food
had moved away. Correct walking around the screen was soon replaced, however, by
walking around on one side only. At the same time all the signs of disturbance of normal
behavior disappeared.

Birds of the crow family make use of information regarding the direction of movement of the stimulus behind the screen in their search for the food. They hardly require individual experience to be able to walk around the screen on the side of movement of the stimulus.

When the problem becomes more complicated (a simple or complex attachment is added to the screen), difficulties are observed in its solution. A definite tendency to walk around the screen on any one side only, irrespective of the direction of movement of the stimulus, is observed. After a little individual experience, however, many birds of the crow family are able to walk around the screen with the simple or complex attachment on the side toward which the stimulus moves, but this problem presents considerable difficulty to their nervous system: some individuals developed a neurotic state, taking the form of fear of the experimental situation.

This investigation demonstrated the existence of a definite relationship between conditioned and extrapolation reflexes. In order to be able to walk around the screen on the side toward which the food moved, most of these animals required individual experience. Only the birds of the crow family were capable of walking around the screen on the side toward which the food moved without preliminary individual experience. When the experimental conditions were made more complicated (the simple or complex attachment was added to the screen) these birds too (although not all of them) required individual experience.

The very great importance of individual experience was revealed in the experiments with fowls. Some of these birds were even able to walk around a screen with a complex attachment on the side of movement of the food, but in order to do this they needed to have individual experience lasting several weeks. In those animals, however, in which extrapolation reflexes were extremely ill defined (for example, in pigeons* and ducks), the addition of prolonged individual experience did not lead to the ability to walk around the screen only on the side toward which the bowl containing the food moved.

The conditioned-reflex "overgrowth" of extrapolation reflexes was clearly seen in the experiments on those of our fowls which,

*After this book was written Fless and Parfenov carried out experiments on four pigeons. In spite of the simplification of the experimental conditions with the screen without attachments (inadmission of the formation of unilateral conditioned-reflex walking movements and shortening of the screen), it was impossible in these birds to achieve systematic walking around the screen on the side of movement of the food.

after prolonged individual experience, were able to walk around the screen with the complex attachment on the side of movement of the food. Simplification of the stereotyped task (opening of the two doors in the complex attachment) led to an increase in the number of mistakes. This change in the experimental conditions, however, did not make it more difficult for the birds of the crow family to solve the problem correctly. These experiments show that the conditioned-reflex components automatize the solution of the task to a greater degree in those animals in which the extrapolation reflexes are relatively weak. When they are well developed, the animal does not require any great individual experience in order to solve the problem adequately, and its solution is automatized by conditioned-reflex components to a lesser degree.

In certain conditions extrapolation and conditioned reflexes may be antagonistic to each other, i.e., negative-induction relationships may be observed between them. Animals with weak extrapolation reflexes in the course of the first, and sometimes of the second, experiment with the screen tried to pass around it on the side toward which the feeding bowl with the food moved. As soon as they developed a unilateral conditioned-reflex walk around the screen, however, these attempts to go in the direction of movement of the bowl containing the food disappeared. After extinction of the unilateral walking around the screen, or after an interval in the work (as a result of which the conditioned-reflex connection was weakened), the extrapolation reflexes began to appear again: the animal tried to walk around the screen on the side toward which the bowl containing the food was moved.

Inhibition of some reflexes by others also clearly emerged when the task was made more complicated. If an animal which had passed around the screen, or even the screen with the simple attachment, on the side of movement of the feeding bowl containing food was not capable of doing so in the experiment with the screen with the complex attachment, it changed over to walking around on one side only. This unilateral walking around was often preserved, however, when the problem was simplified—when the experiments with the screen with the simple attachment, or even with the screen alone, were resumed. The nervous system, having changed from an extrapolation- to a conditioned-reflex method of solving the problem, continued to use this method in the solution of the simpler problems which it had previously tackled by means of extrapolation reflexes.

It is evident that the change to a conditioned-reflex method of solution of the difficult problem is of biological importance. The overstraining of the nervous system during the utilization of extrapolation reflexes leads to the development of a neurotic state, which takes the form of fear of the experimental situation. In our experiments no phobia developed if the animal went around the screen on one side only, i.e., if it solved the problem presented to it by a conditioned-reflex method.

From the discovery of the considerable difference in the degree of development of the extrapolation reflexes in the species of animals studied, it may be suggested that this form of reflex activity undergoes a significant modification in phylogenesis. Investigations of the comparative physiology of conditioned reflex activity, especially those carried out in detail by Voronin (1957) and his co-workers, have shown that the main difference between different classes of animals is revealed not by the velocity of formation of simple conditioned-reflex connections, but by the formation of different forms of internal inhibition and by the development of analyticosynthetic activity.

It may be postulated that the evolution of the higher nervous activity took place not only along the path of modification of unconditioned and conditioned reflexes, but also along the path of development of extrapolation reflexes, which are manifestations of the ability of animals to establish rapidly formed associations between the phenomena of the outside world, standing in a cause-and-effect relationship to each other.

Extrapolation reflexes, which evidently developed at relatively late stages of phylogenesis of vertebrate animals, have widened the adaptive powers of nervous activity, and have made possible adequate "prognostic" reflex reactions during diverse changes in the conditions of the external world.

When discussing the statement that extrapolation reflexes enable the animal to "foresee" the occurrence of elementary events in the future, it must be mentioned that this may come about by means other than extrapolation reflexes.

The prognosis of future events may also be brought about by means of complex unconditioned-reflex activity. A reflex reaction is often effected before the stimulus has come into direct contact with the animal. The distant perception of manifestations of an unconditioned-reflex stimulus "warns" the animal of its impending action, and enables the animal to react adequately to this stimulus. When the shadow of a bird of prey appears over a newly hatched young partridge, the latter falls to the ground or hides. This is the reaction of "foreseeing" the possibility of a future event, the power to do which developed as a result of prolonged natural selection in preceding generations. The unconditioned-reflex "foreseeing" of a future event is a most important condition of an adequate reaction to those comparatively few stimuli in the external environment which are of constant biological importance to the animal.

It is also possible to "foresee" future events by means of conditioned reflexes. Indifferent stimuli, coinciding in time with the action of unconditioned-reflex stimuli, acquire the meaning of positive signals. This takes place, however, only when the relationships between the action of the indifferent and the unconditioned-reflex stimuli are relatively stable. However variable the external environment, the animal must be able to react adequately to the numerous changes in the relationships between stimuli which do not bear a stable relationship to each other. The bird of prey may pounce on its victim from different distances, at different velocities, and from different sides. As a result of the rapidly formed associations between the bird of prey (comprising both the stimulus and the three variables mentioned above) and the stationary environment, the hunted animal is able to extrapolate the direction of movement of its pursuer, and thereby to diminish its chance of being caught.

The use by animals of the ability to extrapolate in order to avoid contact with stimuli evoking a defensive reaction has also been observed in nature. We give a few examples:

Case 1. A truck is moving in a direction perpendicular to that in which a dog is running. If the dog continues to move at the same speed, in a few seconds it will be under the wheels of the truck. The dog, however, quickens its pace and runs past the point of intersection of the two paths before the truck reaches it. As soon as the dog has passed the point of intersection, it again slows its speed and runs at its previous rate.

Case 2. A crow is running across a street along which an automobile is traveling (velocity 80 km/hr). When the vehicle has come to within 30-40 paces, the crow, whose speed of movement is obviously inadequate for it to run across the street in front of the automobile, makes a few beats of its wings and flies across the street in the same direction, alighting on the edge of the road; the automobile passes the bird at a distance of 1.5-2 meters.

It thus becomes possible by means of extrapolation reflexes, brought about as a result of rapidly formed associations between stimuli and the external environment, to make an elementary forecast of future events not only in relatively stable conditions to which the animal is accustomed, but also in an environment in which the relationships between the stimuli are constantly changing.

The question arises whether there are any elementary anatomophysiological mechanisms which lie at the basis of extrapolation reflexes. We consider that some reflections may be given on this matter by way of a working hypothesis.

Our experimental findings suggest that the principal difference between the animal species studied consists, first, of the difference in the duration of traces of acting stimuli in their nervous system, and second, in the utilization of these traces for the formation of rapidly formed associations and, on the basis of these, of extrapolation of the direction of the moving stimulus.

Modern cytological investigations indicate the presence of neuronal mechanisms which retain the traces of acting stimuli. Lorente de Nó (1951) demonstrated the presence of chains of closed rings of neuronal connections in the cerebral cortex, as a result of which not only are impulses of excitation conducted throughout the central nervous system, but, as a result of the circulation of impulses around these closed rings of return connections, the state of excitation of the neurons of the brain is maintained after cessation of the action of the stimulus. We believe that this physiological mechanism of a transient, "operative" memory may be used by the nervous system to retain the trace of a stimulus once it has acted, which is an essential condition for the functioning of extrapolation reflexes.*

This type of memory does not require repeated previous experience, as is necessary for other types of memory, for example long memory, which in Eccles's opinion (1953, 1957) is associated with morphological changes (swelling) in the synapse and the presynaptic nerve fiber, or, according to Morrell (1960), with the

*The importance of the preservation of the effects of stimuli perceived in the past to the perception of spatial images was stressed by Uttley in the article "Temporal and spatial patterns in a conditioned probability machine," in Automata Studies, edited by C. E., Shannon and J. McCarthy, Princeton University Press, Princeton, 1956.

accumulation of ribonucleic acid in the neurons of the brain, and which is the basis of learning.

The work of Polyakov (1956, 1958a, 1958b) showed that the phylogenetically youngest upper layers of the cerebral cortex, which show the most advanced development in the primates and in man, possess, in addition to more complex and perfected neurons, a relatively large number of neurons with short, branching axons (stellate neurons), which form a system of closed rings of return connections. These neurons may also act as the morphological basis of the mechanism providing for retention of traces of previous stimuli. We consider it a likely hypothesis that the evolution of the capacity for an "operative" memory, retaining traces of stimuli which act once only, and the increasing morphological complexity of the system of neuronal rings of return connections took place parallel with each other.

Another aspect of the physiological mechanism of extrapolation is the admission of the "inertia" of the movement in space of the process of stimulation along the neurons of the higher divisions of the brain.* The spatial perception of the relationships existing objectively between things in the outside world is evidently possible only when this relationship is reflected in the neuronal structures of the brain, and therefore when a stimulus moves in space, there must undoubtedly be a change in the relationship between the focus of excitation caused by the moving stimulus and the foci of excitation caused by stationary points in space. It may be admitted that the focus of excitation caused by the action of a moving stimulus will spread not only along those neurons which reflect the spatial relationship of the moving stimulus at a given moment, but also along those which would be in a state of excitation after a certain interval of time if the perception of the moving stimulus were prolonged.

This process may be represented schematically as follows. During the regular movement of the stimulus A, in every small interval of time a continuous, successive series of foci of excitation a will develop in the higher divisions of the central nervous system, which are the projections of the change in the position of the stimulus in space. We admit that the commencing movement of the focus of excitation along the neurons of the brain will continue after the

_____

*Inertia of the spatial movement of the process of stimulation is not, of course, identical with that property of the nervous system, also called inertia, which is characterized by a slow rate of change from a state of excitation into one of inhibition, and vice versa.

stimulus A has ceased, as a result of the action of the obstacle B,*
to have a direct action on the animal's receptor apparatus. The
value m, by which the focus of excitation a is displaced, and which
also is dependent on the time of retention of the focus of excitation
after the obstacle has been brought into action, will also characterize the physiological element of extrapolation. The reflex response R_{n+m} in this case functions, although the animal (as a result
of the obstacle B) did not perceive the movement of the stimulus
A directly with its receptor apparatus in the position N+M.

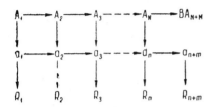

Regular displacement of stimulus A in space

Displacement of the focus of excitation along the neurons of the brain as the projection of the change in the position of A in space

Reflex response of the animal

Under the term "inertia" of the process of stimulation we do
not imply any particular physiological phenomenon. It is evident
that it is based on radiation of the process of stimulation, which
has only a definite trend. The direction of movement of the excitation is determined by the association which is established between
the direction of the stimulus and the environment in which the
movement occurs. The association is established during that interval
of time in which movement of the stimulus A from position 1 to
position N is perceived by the animal's receptors, and its subsequent
course—extrapolation—is possible while the associative connection
which arises is retained by the mechanism of the "operative"
memory.

As these experiments showed, different animals are capable
of forming associations of different degrees of complexity. In
pigeons, in response to movement of the stimulus A from position
1 to position N, an association with the path 1—N is formed. As
a rule, after the stimulus has disappeared behind the obstacle
B, these birds go back along this path (experiments with the corridor and platform). In ducks, fowls, and rabbits an association is
formed between the stimulus A and the obstacle B behind which it
was displaced. These animals seek food mainly near the obstacle B.

*The obstacle B may be a screen, corridor, or the box covering the food moving along
the platform in our experiments, i.e., any condition which, at a certain moment, inter-
feres with perception of the movement of the stimulus A.

In birds of the crow family, cats, and dogs,* an association is formed by the stimulus A not only with the obstacle B but also with the direction of movement of the stimulus, so that they seek the stimulus not only at the place from which it disappears behind the obstacle B, but also in the direction of its movement from N to M.

According to preliminary findings (Polyakov) the comparative study of the cerebral hemispheres in pigeons and birds of the crow family (ravens, crows) has shown that the latter possess an appreciably higher degree of development of the cerebral hemispheres and, which is particularly interesting, of structural differentiation of the neurons than the former. In the corpus striatum of birds of the crow family many neurons were found with short axons, characterized by a much greater variety of shapes and degree of branching of dendrites and axons than in pigeons. There is reason to suppose that these morphological differences play an important part in the difference in capacity for extrapolation which we found between pigeons and birds of the crow family.

This question of the capacities of animals for extrapolation is part of more general questions, namely the formation of associations between stimuli bearing a cause-and-effect relationship to one another, and the ability of the nervous system to "transfer" the individual experience acquired in some conditions to others. In the case of extrapolation we also have a peculiar transfer of individual experience from one situation to another. Some animals are able to use information regarding the movement of a stimulus which they obtain before the obstacle prevents further perception of it, and after the stimulus has ceased to be directly perceived by the receptors. This, of course, is also the transfer of individual experience, but the experience is extremely short. The use of such short individual experience for an adequate reflex reaction is evidently brought about by means of rapidly formed associations between the direction of movement of the stimulus and the external environment.

The experimental results described in this chapter have shown that in the complex and varied behavior of animals it is possible to identify certain relatively elementary reactions of behavior—extrapolation reflexes—which we consider to be essential elements of rational activity. These reflexes may be evaluated and compared in accordance with definite quantitative indices, and they may also

---

*After this book was written, we completed experiments which showed that cats and dogs possess well-developed extrapolation reflexes.

be distinguished from other reflex reactions of the higher divisions of the nervous system, so that it is possible to study the interaction of extrapolation reflexes with other forms of reflex activity, and primarily with conditioned-reflex activity.

The investigation which has been described shows that the quantitative relationship between the degree of expression of the extrapolation and conditioned reflexes is an important condition in defining the path of formation of a behavioral act of an animal which is becoming adapted to the varied and rapidly changing conditions of the outside world.

CONCLUSION

Three principal ideas have been developed in this book. The first is that the basis of behavior is reflex, i.e., determinative, and that it is possible, by means of elementary physiological phenomena, or reflexes, integrated into more complex units, to explain the various aspects of animal behavior, including those which must be regarded as elementary rational forms.

The second idea is that the fundamental properties of the nervous system, and primarily its excitability and strength, are highly important factors in the formation of animal behavior, both normal and pathological.

The third idea is that an overexcited state of the brain, as a result of overloading with afferent impulses, may be the cause of many severe pathological states, some of which may terminate in death.

The study of behavioral acts more complex than the elementary reflexes usually investigated in laboratory conditions showed that their formation takes place as a result of the very close interaction between innate and individually acquired components of behavior. The relationship between the one and the other, however, is not always strictly determined. The same behavioral acts, which we call unitary reactions, and their more complex formations, known as biological forms of behavior, may develop under the predominating influence of both individually acquired and of innate components. Between these extreme paths there exists a continuous series of intermediate forms in which the results of the influence of external and internal conditions are intermingled in different proportions. Our parallel study of the role of hereditary and individually acquired components in the formation of unitary reactions and biological forms of behavior showed that we may hardly speak of the inheritance of a behavioral act. It is more proper to speak of the inheritance of only its unconditioned-reflex component, and of the greater or smaller role of innate factors in the formation of a particular behavioral act.

238   Conclusion

The isolation of certain definite levels of integration of the reflex activity of the nervous system—of unitary reactions and biological forms of behavior—may facilitate the examination of some aspects of the evolution of behavior (Krushinskii, 1944, 1948).

The general tendency in evolution is certainly directed toward the achievement of maximum plasticity and flexibility of the stereotyped behavioral act, adapting the animal to the varied conditions which it may encounter at any moment of its existence.

The diametrically opposite process, i.e., an increase in the relative proportion of innate components in behavioral reactions, initially formed predominantly under the influence of the individual experience of the animal, undoubtedly also takes place in evolution. This must occur in those cases in which in the course of many generations a definite, stereotyped habit is formed in the members of a certain species, which is of permanent biological importance to that species. The selection inevitably commencing under these circumstances, by the stability and rapidity of its formation leads to a relative increase in the proportion of innate reflexes in the development of this habit, which will appear outwardly as the hereditary "fixation" of an individually acquired habit, and its gradual conversion into an innate, instinctive behavioral act.*

The astounding "expediency," the adaptability of innate instinctive behavioral acts to the conditions in which the animal lives, becomes comprehensible if it is regarded as evolving along a path laid down by the animal's individual experience. All the many "trials and errors" by means of which every animal is best adapted to conditions of life, thereby laying open the inherited specificity of its nervous activity to the action of selection, may thereby exert a formative influence also on the behavior of the offspring. Thus although there is a profound qualitative difference between individually acquired and inborn behavioral acts, the conversion of inborn behavior into individually acquired acts, and vice versa, may proceed along the lines of purely quantitative changes. The possibility of evolution of the ratio of inborn and individually acquired components taking part in the formation of a unitary reaction shows how necessary it is to admit the presence of continuous transitions between innate and individually acquired behavior.

*An illustration of this statement is an example of the "carriage" behavior of dogs, described by Keller and Trimble (1940) and mentioned above, when prolonged selection among Dalmatian hounds, directed toward ability to be trained in a specific behavioral act—to run with a carriage—led to the result that modern individuals of this breed have an innate desire to run with a carriage.

Our experimental analysis of defensive and retrieving behavior in dogs showed the ways by which behavioral acts are formed in animals and demonstrated the existence of such a continuity.*

The introduction of the concept of the unitary reaction thus not only makes it possible to define the elementary behavioral act of an animal and to give a general scheme of the relative role of innate and individually acquired components in its formation, but it may also aid in the elucidation of some of the problems of the evolutionary and comparative physiology of the nervous system, which have been developed successfully in the USSR by Orbeli (1945, 1958), Koshtoyants (1951, 1957), Voronin (1958), Biryukov (1958), Slonim (1958), and others, and which are becoming increasingly important at the present time.

The questions discussed in this book are directly concerned with the problem of instinct.

Darwin gave, for his time, a clear definition of instinct: "An action, which we ourselves require experience to enable us to perform, when performed by an animal, more especially by a very young one, without experience, and when performed by many individuals in the same way, without their knowing for what purpose it is performed, is usually said to be instinctive."†

Pavlov pointed out that instincts are not a special property of animal behavior, but consist of complex chains of inborn, unconditioned reflexes. This view of the unconditioned-reflex nature of instinct, however, while formulated perfectly correctly by Pavlov, does not exhaust the whole complexity of the phenomenon. The chief difficulty in the definition of instinctive behavioral acts is that it is practically impossible to draw a line between behavioral acts formed with the predominant participation of innate, unconditioned-reflex or of individually-acquired components.

*This may also be well illustrated by the formation of the reaction of young birds toward their mother. By his studies of the formation of behavior in young birds, Lorenz (1935) showed that the reaction of the young quacking duck to its mother is fixed strictly by inheritance. The principal specific sign of the quacking duck, to which (and only to which) the duckling reacts is its manner of quacking. It is quite a different matter with the reaction of the grey gosling to its mother. Its behavioral reaction is nonspecific and is due only to individual experience. If hatched out under any other bird it will "recognize" this bird as its mother. Not only a bird, but any moving object which first happens to catch the eye of the newly hatched gosling will be looked upon as its "mother," and it will no longer pay attention to its real parents. Between these extreme cases of the formation of such behavior in different species of birds there is a continuous series of intermediate forms in which innate and individually acquired components of behavior are present in different proportions.

†C. Darwin. "Instinct and notes on instinct." Collected Works [Russian translation] (Biomedgiz, p. 160), Vol. 3, 1938. (Origin of Species, Chapter 8.)

If, for example, we examine from this point of view a behavioral act such as that of a retriever dog standing over game, it is difficult to define this as instinct. Innate unconditioned-reflex components naturally play an extremely important role in the performance of this act. Darwin, for instance, pointed out that puppies often point without any previous training. Nevertheless, every pointer must have individual experience (which differs greatly in different dogs) before it will begin to seek and point reliably to game of a particular species.

There are, of course, behavioral acts in the formation of which individual experience plays practically no part. Darwin rightly pointed out that "no one would ever have thought of teaching, or probably could have taught, the tumbler-pigeon to tumble, an action which, as I have witnessed, is performed by young birds that have never seen a pigeon tumble."

We consider that as a result of this analysis we may regard instincts not simply as innate unconditioned-reflex behavior, but as behavioral acts formed with a high relative proportion of innate components, although usually with the participation of individual experience. Under these circumstances the conditioned-reflex component adapts the animal's behavior to the concrete conditions of its existence, and thereby modifies the manifestation of the innate, unconditioned-reflex components of behavior.

The unitary reactions and biological forms of behavior, in the formation of which innate unconditioned-reflex components predominate, may thus be defined as instinctive behavioral acts. Unitary reactions and their more complex units — biological forms of behavior, in which individually acquired components predominate, are evidently behavioral acts which it is customary to call habits.

Consequently, between the instincts and the habit, each of which is formed as a result of the interaction between individually acquired and innate reflexes, there is a continuous series of intermediate forms, and the main criterion of their difference is the relative proportion of conditioned and unconditioned reflexes taking part in their formation. We hope that this analysis will assist in the clarification of one aspect of that problem of such importance to the whole subject of behavior, namely what is instinct.

During our study of the behavior of animals we have shown that increased excitability of the nervous system is an extremely favorable basis upon which a series of individually-acquired habits may most easily be formed, and individual acts of animal behavior most clearly manifested. A change in the excitability of the nervous system associated with endocrine imbalance (hyper- or hypothyroidism, for example) may also influence the formation of an animal's behavior.

Another functional property of the nervous system, determining to a large extent the behavioral reactions of an animal, is the strength of the nervous system. The importance of individual differences in this property of nervous activity is especially apparent when the animal is called upon to make a considerable physiological effort. In these conditions the animal with a strong nervous system, because of the quicker formation of individually acquired habits, adapts itself more rapidly to the external environment. The strength of the nervous system is also dependent on the gonads. Male sex hormone raises the limit of functional capacity of the nervous system.

Our findings demonstrate the importance of the fundamental typological properties of the nervous system, as distinguished by Pavlov, in the behavior of dogs, both under laboratory conditions and in the varied conditions of life.

It should be mentioned that we (Krushinskii, 1946d, 1946e) have established that some relationship exists between the constitution and physique of dogs and the typological properties of their nervous activity.

In accordance with their constitution and physique, we have subdivided dogs along the lines of Malinovskii's (1935, 1945) scheme for man, adapting it to suit these animals. An investigation on a large number of dogs (243 animals), the results of which were treated statistically, provided a basis for the objective evaluation of the connections existing between the different types of physique and the behavioral properties of dogs. In general terms it can be said that "athletic" and "wide-bodied" types of physique in dogs are associated with a tendency toward greater strength and lower excitability of the nervous system. Conversely, a "slender" and "narrow-bodied" physique in dogs is associated with a tendency toward increased excitability and weakness of the nervous system. Bearing in mind that the conditions of formation of the bodily structure in man and the dog are very different, it is nevertheless interesting to note that Roginskii (1937) found greater weakness

Conclusion

of the nervous system associated with the cerebral type of physique in man (corresponding to the "slender" type in dogs) than with other types of physique.

The degree of excitability and the strength of the nervous system, which have an important influence on the formation of the normal behavior of animals, are also very important factors in the development of various pathological disturbances of nervous activity.

These investigations showed that in rats with increased excitability and weakness of inhibition, when a strong sound stimulus was applied, a severe excitation of the brain developed, which could have very serious consequences: a convulsive epileptiform fit, a cataleptoid state, obsessive tic-like hyperkinesis of the type of myoclonic spasms, and acute disturbances of the cerebral circulation with atony of the capillaries and hemorrhages.

It has been shown that the nervous system has a remarkable mechanism of double assurance in its struggle against this pathological excitation. Active inhibition, interrupting the beginning excitation, is the "first line of defense." This inhibition, however, is not stable and is soon exhausted, and is replaced by limiting parabiotic inhibition, the "second line of defense" of the nervous system, which not only prevents a further increase in the excitability of the nerve cells, but also actively lowers it. The increasing excitation of the brain, leading to deepening of the limiting, parabiotic inhibition, thereby lowers its excitation. From the investigations of Krushinskii, Korzhov, and Molodkina (1958) there is reason to suppose that the therapeutic effect of electric shock may be based upon the same mechanism. Electric shock, by causing severe excitation of the brain, strengthens parabiotic inhibition, which leads to a decrease in the excitability of the brain.

We consider that the severe excitation with prolonged inertia, developing under the influence of external stimuli, is identical with that phenomenon which was described by Vvedenskii (1912) under the term "hysteriosis," and is an important physiological mechanism lying at the basis of many disturbances of nervous activity.

Severe excitation of the neurons of the brain under the influence of an excessive number of afferent impulses may lead to the development of shock, with signs of acute disturbance of the circulation and with fatal cerebral hemorrhage. The role of primary excitation of the brain in the genesis of shock is mentioned by Koreisha (1955, 1957), who showed that as a result of excessively strong stimulation of the cerebral hemispheres, irrespective of the site of application of the stimulating agent, a general disturbance of the circulation developed, beginning with spasm of the vessels and an increase in blood pressure, followed by dilatation of the capillaries and pre-capillaries. The permeability of the vessel walls is disturbed and hemorrhages take place as a result of diapedesis.

It has often been stated in the literature that shock is based upon limiting inhibition, extending not only to the cortex but also to the brainstem, with its vasomotor centers. Petrov (1956), for example, writes that in shock "limiting inhibition arises at first in the cerebral cortex and subsequently radiates to the subcortical divisions of the brain, and in particular it develops in the vaso-motor centers."

Our experimental findings indicate rather that, on the contrary, so long as there is limiting, protective inhibition, shock does not develop. Shock develops when, during a state of excitation, the defensive inhibitory functions of the nervous system are inadequate to withstand the severe excitation of the neurons. The state of inhibition which is observed during shock must evidently be re-garded as a secondary phenomenon, arising as a result of the acute disturbance of the cerebral circulation, the fall of blood pressure, the lowering of the body temperature, and other severe functional disturbances.

The problem of the increased vascular permeability and of hemorrhage in neurogenic shock is undoubtedly one of great prac-tical concern, for hemorrhage in various organs often complicates the picture of shock and is justifiably regarded (Akhutin, 1945) as a sequela of nerve shock.

When the importance of nervous trauma in the development of various pathological conditions is being discussed, its role must not be exaggerated, and it must not be regarded as a possible cause of the development of any pathological process. For instance, we investigated the influence of an excited state of the brain on the development of malignant tumors (Krushinskii, Molodkina, Prigo-zhina, and Shabad, 1954). This investigation showed that induced tumors (as carcinogen we used 9,10-dimethyl-1,2-benzanthracene) develop equally often in rats traumatized by means of a sound stimulus and in nontraumatized rats.

The results obtained from the study of the physiological me-chanisms of the state of excitation caused by the action of a sound stimulus may be relevant to the solution of a number of practical problems.

In 1952, for the prevention of epileptic fits in clinical practice we began to use a preparation suggested by Sereiskii. The physiological study of the anticonvulsant action of this preparation, which we carried out on an experimental model, namely reflex epilepsy in rats, showed that the presence of caffeine in the mixture lowered the efficacy of the preparation in the treatment of epilepsy (Krushinskii, Sereiskii, Pushkarskaya, and Fedorova, 1955). This was to be expected, for caffeine, by increasing the excitability of the nervous system, was bound to increase the intensity of the convulsive fits. In accordance with this effect, the dose of caffeine in the recommended preparation for treatment of epilepsy was greatly reduced (to 0.02–0.15 g),* and it was advised that it be clinically prescribed in this way. Investigations are going on in our laboratory at the present time to strengthen the action of this particular preparation for the treatment of epilepsy.

Investigations which we carried out (Krushinskii, Pushkarskaya, and Molodkina, 1953) also indicate that care must be used when giving large doses of caffeine, which may increase the severity of neurogenic shock.

Another example, which we believe to be of definite practical interest, is the investigation which we carried out (Krushinskii and Molodkina, 1949, 1957) showing that spinal injuries with subsequent hemorrhages (which frequently follow shock therapy) in the course of a convulsive fit in rats are the direct consequence of a deficiency of calcium in the diet.

The necessity for care in the administration of large doses of vitamins C and P is also demonstrated by the researches of Fless, who showed that the administration of these vitamins in doses of 30–120 mg/kg daily for 2–3 weeks not only did not protect rats against the development of shock or hemorrhage after prolonged exposure to a sound stimulus but, on the contrary, caused a considerable increase in the proportion of fatal cases.

In a laboratory of acoustic physiology in France, Lehmann and Busnel (1959) successfully used sound-induced convulsive fits in mice to test various neuroplegic and sedative drugs (Busnel, Lehmann, and Busnel, 1958). Besides other model experiments in animals, these investigations were responsible for the introduction of a number of new and effective drugs for the treatment of nervous

---

*Sereiskii considered that it was irrational to exclude caffeine completely, and believed that a small dose of it in the preparation was necessary in order to dispel the hypnotic state developing in the patient as a result of the administration of bromural and luminal, which are also included in the preparation.

and mental diseases (meprobamate, methylpentinol, benactizine, etc.). On the other hand, it was also shown that reserpine, widely used in clinical practice as an effective tranquilizer, intensifies convulsive fits caused by a sound stimulus, by an electric current, and by administration of camphor and cardiazol. Death was observed in some cases after combined treatment with reserpine and electric convulsion therapy.

Preliminary trials of the action of various drugs on fits of reflex (acoustic) epilepsy in rats and mice are thus one of the methods which may assist in elucidating the comparative efficacy of these drugs and, in some cases, show that care must be taken when they are used.

In conclusion let us turn to the question of the different levels of integration of the complex forms of behavior. As a result of our investigation of animal behavior we were able to distinguish reflexes which we called extrapolation reflexes. These are evidently formed on the basis of rapidly developing associations between the direction of movement of a stimulus and the external environment, and they undoubtedly have a most important part to play in the formation of that level of animal behavior which can be defined as the simplest form of rational behavior.

We consider "intellectual" or "rational" activity to be that reflex activity of the higher divisions of the nervous system which comprises in the simplest case at least the following acts.

1. The identification from among the whole variety of stimuli in the external environment of those which, at a given moment, are of definite biological importance to the animal.

2. The rapid formation of associations between stimuli bearing cause-and-effect relationships to each other.

3. The extrapolation of subsequent changes in the relationships between the stimuli on the basis of the associations thus formed.

This operation by means of reflections of concrete phenomena of the outside world as a change takes place in the relationships between them, and the extrapolation of the direction of this change must increase the adaptive powers of the animal.

Extrapolation reflexes undoubtedly may act as components of integrated groups of reflexes, namely unitary reactions and biological forms of behavior, thereby increasing their adaptive importance. Our investigation shows that animals possess the ability to establish certain cause-and-effect relationships between the stimuli of the outside world, and they use the information obtained in adaptation to the varied and constantly changing conditions of the external environment. This suggests that they possess the rudiments of rational activity, which are evidently the physiological basis on which the human intellect was formed during subsequent evolution under the influence of social and work relationships.

The power of establishing cause-and-effect relationships between the phenomena of the outside world, so exceptionally well developed in man, developed, it must be considered, from that form of association in animals which lies at the basis of the detection of the most elementary cause-and-effect relationships.

We began from the notion that the rational activity of animals is based upon purely physiological processes, entirely amenable to objective study. We hope that our analysis of the formation of animal behavior will confirm yet again that even complex integrated forms of behavior are ultimately produced by the reflex principle.

As Sechenov pointed out, "the question whether all mental activities are reflex in type or not will be answered in the affirmative from the general point of view if it can be proved that the initial forms from which the whole psychic life grows consist of acts performed in this manner, and that the nature of the processes is not changed throughout all subsequent phases of mental development." *

---

*I. M. Sechenov, Selected Works (Izd. VIEM, Moscow, 1935), p. 258.

BIBLIOGRAPHY

SOVIET LITERATURE

Adamez, L., General Zootechnics. Sel'khozgiz, Moscow–Leningrad, 1930.

Akhutin, M. N., Foreword. Transactions of Group No. 1 on the Study of Shock, 1945.

Akimov, N. E., The importance of sex differentiation in the acquisition of habits by white rats. Problemy Sovremennoi Psikhol., Vol. 3 (1928).

Akimov. N. E., The importance of sex differentiation in the acquisition of habits by rats. Problemy Sovremennoi Psikhol., Vol. 6 (1930).

Al'bitskii P. A., The return action or 'after-effect' of carbon dioxide and the biological action of CO_2 usually present in the body. Izvest. Voenno-Med. Akad., Vol. 22-23 (1911).

Alekseeva I. A. Conditioned reflexes to a multicomponent chain stimulus in a dog in conditions of free motor activity. Zhur. Vysshei Nerv. Deyatel, Vol. 6. No. 4 (1956).

Alekseeva, T. T., A case of persistent change in the character of the conditioned motor reaction of dogs in conditions of active-motor selection. In: Problems of Higher Nervous Activity (edited by P. K. Anokhin). Izd. AMN SSR, Moscow, 1949.

Andreev, L. A., Manual on the Training of Specialists in Dog Breeding for the Services. Sel'khozgiz, Moscow, 1939.

Andreev, L. A. and Pugsley, L., The effect of the hormone of the parathyroid glands and of vitamin D on the higher nervous activity of dogs. Fiziol. Zhur. SSSR, Vol. 18, No. 1 (1935).

Anichkov, N. N., The study of diseases by creation of their experimental models. General Assembly of the Academy of Sciences of the USSR on the Thirtieth Anniversary of the Great October Socialist Revolution. Moscow–Leningrad, 1948.

Anokhin, P. K., The study of the dynamics of higher nervous activity. I. The active secretory-motor method of study of higher nervous activity. Nizhegorod. Med. Zhur., No. 7-8 (1932).

Anokhin, P. K., The problem of the center and the periphery in the modern physiology of nervous activity. In: The Problem of the Center and Periphery in the Physiology of Nervous Activity. Gor'kii, 1935.

Anokhin, P. K., From Descartes to Pavlov. Medgiz, Moscow, 1945.

Anokhin, P. K., The physiological substrate of signal reactions. Zhur. Vysshei Nerv. Deyatel., Vol. 7, No. 1 (1957).

Anokhin, P. K. and Strezh, E., The study of the dynamics of higher nervous activity. III. Disturbance of the active selection as a result of replacement of the unconditioned stimulus. Fiziol. Zhur. SSSR, Vol. 16, No. 2 (1933).

Anokhin, P. K. and Strezh, E., The study of the dynamics of higher nervous activity. VI. Characteristics of the receptive function of the cerebral cortex at the moment of action of the unconditioned stimulus. Fiziol. Zhur. SSSR, Vol. 17, No. 6 (1934).

Anokhin, P. K., Shumilina, N., Anokhina, A., et al., The functional system as the basis of integration of nervous processes in embryogenesis. Collected Proceedings of the Sixth All-Union Congress of Physiologists, Biochemists and Pharmacologists, 1937.

Apter, I. M., Experimental Neuroses in the Motor Sphere in Dogs in Experiments in the Laboratory and in Natural Conditions. Doctorate dissertation, Khar'kov, 1952.

Arkhangel'skii, V. M., The effect of transplantation of the testicles on the higher conditioned-reflex nervous activity of dogs. Dnepropetrovsk. Med. Zhur., No. 5-6 (1927).

Arkhangel'skii, V. M., Conditioned reflexes in castrated dogs. Russk. Fiziol. Zhur., Vol. 14, No. 2-3 (1931).

Arkhangel'skii, V. M., The effect of hormones of the sex glands on the work of the cerebral cortex in dogs. Nauch. Zap. Dnepropetrovsk. Univ., Vol. 1 (1937).

Arkhangel'skii, V. M., The influence of the thyroid gland on excitation and inhibition in the cerebral cortex. Fiziol. Zhur. SSSR, Vol. 30, No. 1 (1941).

Artem'ev, E. I., The effect of castration on the higher nervous activity of female dogs. Trudy Gor'k. Ped. Inst., Vol. 4 (1939).

Ashby, W. R., Design for an intelligence-amplifier. In: Automata Studies [Russian translation], IL, Moscow, 1956. (Original English published as: C. E. Shannon and J. MacCarthy, [ed.], Annals of Mathematical Studies No. 34, Princeton University, Princeton, New Jersey, 1956.)

Azimov, G. I., The conditioned reflex activity of the thyroidectomized animal. Zhur. Eksptl. Biol. i Med., Vol. 8, No. 19 (1927).

Bam, L. A., The effect of calcium chloride on the higher nervous activity of the dog. Fiziol. Zhur. SSSR, Vol. 27, No. 6 (1939).

Baranov, V. G., Speranskaya, E. N., and Tendler, D. S., The effect of small doses of thyroidin on the higher nervous activity of dogs. Abstracts of Proceedings of a Joint Session of the All-Union and Ukrainian Institute of Experimental Endocrinology, 1954.

Baturenko, T. I., Peptone shock associated with functional disturbance of the thyroid gland. Byull. Dnepropetrovsk. Med. Inst., No. 2 (1940).

Baturenko, T. I., Peptone shock associated with functional disturbance of the thyroid gland. Farmakol. i Toksiol., Vol. 4, No. 3 (1941).

Bekhterev, V. M., The General Basis of Human Reflexology. Gosizdat, Moscow–Leningrad, 1928.

Belichenko, A. B., Evidence on the Nature of Formation of Traumatic Shock. Doctorate dissertation, Kursk, 1953.

Beritashvili, I. S., The comparative study of the individual behavior of the higher vertebrate animals. Vestnik Akad. Nauk SSSR, No. 10 (1939).

Beritov, I. S., The Individually Acquired Activity of the Central Nervous System. Tiflis, 1932.

Beritov, I. S., The Principal Forms of Nervous and Psychonervous Activity. Izd. AN SSSR, Moscow–Leningrad, 1947.

Biryukov, D. A., The role and the tasks of comparative physiology and pathology of higher nervous activity in medicine. In: Evolution of the Functions of the Nervous System. Medgiz, Leningrad, 1958.

Bogolepov, N. K., Comatose States (Clinical Aspects and Treatment). Medgiz, Moscow, 1950.

Bogoyavlenskaya, N. V., The clotting system of the blood and cerebral hemorrhage under the influence of nervous trauma. Byull. Eksptl. Biol. i Med., Vol. 44, No. 9 (1957).

Bogoyavlenskaya, N. V., The role of the nervous system in the regulation of the mechanism of coagulation of the blood. Uspekhi Sovremennoi Biol., Vol. 45, No. 1 (1958).

Bolotova, T. I., Disturbance of vascular permeability in Basedow's disease. In: Essays on Vascular Permeability. Medgiz, Moscow, 1956.

Brazier, M., The Electrical Activity of the Nervous System [Russian translation]. IL, Moscow, 1955.

Burdenko, N. N. and Smirnova, L. G., The adrenal syndrone of shock. In: Shock (Transactions of a Conference on the Problem of Shock). Kiev, 1938.

Bures, J., External inhibition of reflex epilepsy in rats and mice. Chekhoslovatskaya Fiziol., Vol. 2, No. 1 (1953).

Bures, J., The influence of exclusion of analyzers on the disposition to convulsions in rats and mice. Chekhoslovatskaya Fiziol., Vol. 2, No. 3 (1953a).

Bures, J., The influence of a general lowering of the body temperature on the disposition of rats and mice to reflex epilepsy. Chekhoslovatskaya Fiziol., Vol. 2, No. 4 (1953b).

Darwin, C., On the Origin of Species by means of Natural Selection, or the Preservation of Favoured Races in the Struggle for Life, translated by K. Timiryazev. St. Petersburg, 1896.

Darwin, C., Instinct, and notes on instinct. In: Collected Works [Russian translation], Biomedgiz, 1938, Vol. 3.

Davidenkov, S. N., Evolutionary and Genetic Problems in Neuropathology. Leningrad, 1947.

Dobrokhotova, L. P., The effect of methylthiouracil on hemorrhage and shock developing under the influence of nervous trauma. Doklady Akad. Nauk SSSR, Vol. 114, No. 6 (1957).

Dobrokhotova, L. P., The influence of hyperthyroidization on the functional state of the central nervous system during the development of states of shock and hemorrhage in animals under the influence of nervous trauma. Problemy Endokrinol. i Gormonoterap., Vol. 4, No. 3 (1958).

Dolin, A. O. and Zborovskaya, I. I., A further study of the role of the process of inhibition (inductive inhibition) in the suppression of pathological states. Zhur. Vysshei Nerv. Deyatel., Vol. 2, No. 1 (1952).

Erofeeva, M. N., Electrical Stimulation of the Skin of the Dog as a Conditioned Stimulus of the Work of the Salivary Glands. Dissertation, St. Petersburg, 1912.

Fless, D. A., The Study of Higher Nervous Activity and Intuition in Tracker Dogs. Candidate dissertation, Moscow, 1952.

Fless, D. A., The influence of factors changing the state of the processes of excitation and inhibition on the phases of reflex epilepsy. Abstracts of Proceedings of a Scientific Conference on Problems of the Experimental Pathophysiology and Therapy of the Higher Nervous Activity of Animals. Moscow, 1957.

Fol'bort, G.V., New facts and observations regarding Pavlov's teaching on higher nervous activity. Zhur. Vysshei Nerv. Deyatel., Vol. 1, No. 3 (1951).

Frolov, Yu. P., The passive-defensive reflex and its sequelae. In: Collection in Honor of Academician I. P. Pavlov's Seventieth Birthday, 1925.

Gekker, E. I., The effect of implantation of the ovaries on the conditioned-reflex activity of female dogs. Kazan. Med. Zhur.No. 4-5 (1931).

Glisson, B. and Shumilina, N., The study of the dynamics of higher nervous activity. I. Reorganization of motor conditioned reflexes with no change in the unconditioned reinforcement. Fiziol. Zhur. SSSR, Vol. 30, No. 4 (1941).

Gunin, V. I., The mechanism of action of aluminum and potassium thiocyanate and of thiouracil on the higher nervous activity and the thyroid gland of the dog. Collection of Proceedings of the Second All-Union Conference of Students' Scientific Societies of the Medical, Pharmaceutical and Stomatological Institutes, 1952.

Gusel'nikova, K. G. and Krushinskaya, N. L., Changes in the bioelectrical activity of certain divisions of the cerebellum and motor area of the cortex during a sound-induced epileptiform fit. Nauch. Dokl. Vysshei Shkoly, Biol. Nauki, No. 2 (1958).

Gusel'nikova, K. G., Data on the mechanism of the sound-induced epileptiform fit in rats. Nauch. Dokl. Vysshei Shkoly, Biol. Nauki, No. 1 (1959).

Il'in, N. A., Dissociation during crossing of wolf and dog. Tr. po Dinamike Razvitiya, Vol. 8 (1934).

Ivanov-Smolenskii, A. G., The passive-defensive reflexes and the strong type of nervous system. Trudy Fiziol. Lab. Akad. I. P. Pavlova, Vol. 4, Nos. 1 and 2 (1932).

Ivanov-Smolenskii, A. G., Ways of development of Pavlov's ideas in the field of the pathophysiology of higher nervous activity. Proceedings of a Scientific Session on the Physiological Teaching of Academician I. P. Pavlov. Izd. AN SSSR, Moscow, 1950.

Kerekesh, F. S., Investigations of the permeability of capillaries in goiter. Nauch. Zap. Uzhgorodsk. Univ., Vol. 15 (1955).

Kleshchov, S. V., Administration of large doses of caffeine to determine the type of the nervous system. Trudy Fiziol. Lab. Akad. I. P. Pavlova, Vol. 8 (1938).

Kleshchov, S. V., Treatment of motor neurosis by external inhibition. Trudy Fiziol. Lab. Akad. I. P. Pavlova, Vol. 8 (1938a).

Köhler, W., Investigation of the Intellect of Anthropoid Monkeys. Izd. Komm. Akad., Moscow, 1930.

Kolesnikov, M. S., Peculiarities of the higher nervous activity of some generations of dogs of weak type. Abstracts of Research during 1947 at the Academy of Medical Sciences of the USSR (Biological Sciences Division), No. 1, 1947.

Kolesnikov, M. S. and Troshikhin, V. A., Small standard of tests to determine the type of higher nervous activity of dogs. Zhur. Vysshei Nerv. Deyatel., Vol. 1, No. 5 (1951).

Kolesnikov, M. S., Short scheme of tests to determine the type of higher nervous activity of dogs. Trudy Inst. Fiziol. AN Belorus. SSR, Vol. 2 (1958).

Kol'tsova, M. P., (see Sadovnikova-Kol'tsova, M. P.)

Komissarenko, V. P., Buiko, S. M., Glauzman, F. A., and Teplitskaya, E. O., The conditioned-reflex activity of dogs with experimental hypothyroidism. Abstracts of Reports on the Mechanism of Action of Hormones. 1957.

Konge, L. A., The influence of thyroidin on the higher nervous activity of dogs. Abstracts of Proceedings of a Conference on the Evolutionary Physiology of the Nervous System, 1956.

Koreisha, L. A., The problem of traumatic shock in the light of modern ideas of the nervous regulation of the circulation of the blood. Abstracts of proceedings of the Sections of the Twenty-Sixth Congress of Surgeons of the USSR. Moscow, 1955.

Koreisha, L. A., The problem of shock after injury to the human brain in the light of modern ideas of the nervous regulation of the circulation of the blood. In: Problems of Modern Neurosurgery. Moscow, 1957, Vol. 2.

Koshtoyants, Kh. S., Fundamentals of Comparative Physiology, 2nd edition, enlarged and revised. Moscow–Leningrad, 1951, Vol. 1.

Koshtoyants, Kh. S., Fundamentals of Comparative Physiology. Izd. AN SSSR, Moscow, 1957, Vol. 2.

Kotlyar, B. I., The role of the cerebral cortex in the development of motor pathological reactions induced in rats by the method of sound stimulation. Nauch. Dokl. Vysshei Shkoly (Biol. Nauki), No. 4 (1958).

Kotlyar, B. I., The localization of clonic convulsions of an epileptic fit. Nauch. Dokl. Vysshei Shkoly (Biol. Nauki), No. 2 (1959).

Krasuskii, V.K., The inheritance of some properties of the higher nervous activity of dogs with a strong type of nervous system. Trudy Inst. Fiziol. im. I. P. Pavlova, Vol. 2 (1953).

Kreps, E.M., The influence of estrus on the higher nervous activity of the dog. Russk. Fiziol. Zhur., Vol. 6, No. 4-6 (1924).

Krushinskii, L.V., Investigation of the phenogenetics of behavioral signs in dogs. Biol. Zhur., Vol. 7, No. 4 (1938).

Krushinskii, L.V., The thyroid gland and the defensive behavioral reactions of dogs. Biol. Zhur., Vol. 7, No. 5-6 (1938a).

Krushinskii, L.V., The inherited 'fixation' of the individually-acquired behavior of animals and the formation of instincts. Zhur. Obshchei Biol., Vol. 5, No. 5 (1944).

Krushinskii, L.V., Relationships between the active and passive defensive reactions of dogs. Izvest. Akad. Nauk SSSR, Ser. Biol., No. 1 (1945).

Krushinskii, L.V., The role of heredity and of conditions of training on the manifestation and expression of behavioral signs in dogs. Invest. Akad. Nauk SSSR, Ser. Biol., No. 1 (1946).

Krushinskii, L.V., The inheritance of behavioral properties in animals. Uspekhi Sovremennoi Biol., Vol. 22, No. 1 (1946a).

Krushinskii, L.V., Sex differences in the behavior of dogs. Zhur. Obshchei Biol., Vol. 7, No. 2 (1946b).

Krushinskii, L.V., The role of genotypic and humoral factors in the behavioral development of rats. Dokaldy Akad. Nauk SSSR, Vol. 52, No. 6 (1946c).

Krushinskii, L.V., Correlation between the constitutional body structure and behavior of of dogs. Doklady Akad. Nauk SSSR, Vol. 52, No. 7 (1946d).

Krushinskii, L.V., Analysis of the Formation of the Behavior of Animals in Ontogenesis. Doctorate dissertation, Moscow, 1946e.

Krushinskii, L.V., The importance of peripheral impulses in the sex form of behavior of females. Doklady Akad. Nauk SSSR, Vol. 55, No. 5 (1947).

Krushinskii, L.V., Inheritance of passive-defensive behavior (cowardice) and its association with the types of nervous system in dogs. Trudy Inst. Evolutsion. Fiziol. i Patol. Vysshei Nerv. Deyatel. im. Akad. I.P. Pavlova (Koltushi), Vol. 1 (1947a).

Krushinskii, L.V., Relationship between inherited and individually-acquired components in the formation of animal behavior. Abstracts of Proceedings of the Seventh All-Union Congress of Physiologists, Biochemists, and Pharmacologists, 1947b.

Krushinskii, L.V., Some stages of integration in the formation of animal behavior. Uspekhi Sovremennoi Biol., Vol. 26, No. 2 (5) (1948).

Krushinskii, L.V., Advances in the study of experimental epilepsy and of the physiological mechanisms underlying it. Uspekhi Sovremennoi Biol., Vol. 28, No. 1 (4) (1949).

Krushinskii, L.V., Section in: The Service Dog. Izd. Sel'khozgiz, Moscow, 1952.

Krushinskii, L.V., Study of the relationship between excitation and inhibition in normal and pathological conditions by the method of sound stimulation. Uspekhi Sovremennoi Biol., Vol. 37, No. 1 (1954).

Krushinskii, L.V., Extrapolation reflexes in birds. Uch. Zap. MGU (Ornithology), No. 197 (1958).

Krushinskii, L.V., The biological importance of extrapolation reflexes in animals. Zhur. Obshchei Biol., Vol. 19, No. 6 (1958).

Krushinskii, L.V., Extrapolation reflexes as the elementary basis of rational activity in animals. Doklady Akad. Nauk SSSR, Vol. 121, No. 4 (1958).

Krushinskii, L.V., Genetic investigations on the experimental pathophysiology of higher nervous activity. Byull. MOIP, Otdel. Biol., Vol. 64, No. 1 (1959).

Krushinskii, L.V., The study of extrapolation reflexes in animals. In: Problems of Cybernetics. Fizmatgiz, Moscow, 1959a, No. 2.

Krushinskii, L.V., Investigation of the physiological mechanisms of the fits in reflex epilepsy. In: The Problem of Epilepsy. Medgiz, 1959b.

Krushinskii, L.V., Chuvaev, A.V., and Volkind, N.Ya., New data concerning the study of of scenting in dogs. Zool. Zhur, Vol. 25, No. 4 (1946).

Krushinskii, L.V. and Kabak, Ya. M., Change in the degree of excitability of the nervous system by means of methylthiouracil. Doklady Akad. Nauk SSSR, Vol. 57, No. 7 (1947).

Krushinskii, L.V. and Molodkina, L.N., Paralyses due to hemorrhage into the central nervous system after experimental epileptic fits in rats. Doklady Akad. Nauk SSSR, Vol. 66, No. 2 (1949).

Krushinskii, L.V., Molodkina, L.N., and Kitsovskaya, I.A., The role of the parathyroid glands in the production of fits in experimental epilepsy. Byull. Eksptl. Biol. i Med., Vol. 30, No. 8 (1950).

Krushinskii, L. V., Fless, D. A., and Molodkina, L. N., Analysis of the physiological proc-
    esses underlying experimental reflex epilepsy. Zhur. Obshchei Biol., Vol. 11, No. 2
    (1950).
Krushinskii, L. V., Fless, D. A., and Molodkina, L. N., The study of limiting inhibition by
    the method of sound stimulation. Byull. Eksptl. Biol. i Med., Vol. 33, No. 4 (1952).
Krushinskii, L. V., Pushkarskaya, L. P., and Molodkina, L. N., An experimental study of
    cerebral hemorrhage under the influence of nervous trauma. Vestnik MGU, No. 12
    (1953).
Krushinskii, L. V., Molodkina, L. N., Prigozhina, E. L., and Shabad, L. M., The study of
    the role of nervous trauma in the development of malignant tumors. Zhur. Vysshei
    Nerv. Deyatel., Vol. 4, No. 6 (1954).
Krushinskii, L. V., Sereiskii, M. Ya., Pushkarskaya, L. P., and Fedorova, G. I., An ex-
    perimental study of a new antiepilepsy preparation. Zhur. Vysshei Nerv. Deyatel.,
    Vol. 5, No. 6 (1955).
Krushinskii, L. V., and Molodkina, L. N., Hemorrhages into the spinal cord as a result of
    fits of experimental epilepsy. Uspekhi Sovremennoi Biol., Vol. 44, No. 2 (5) (1957).
Krushinskii, L. V. and Dobrokhotova, L. P., The influence of the thyroid gland on the
    mortality from shock and hemorrhage caused by strong sound stimuli. Byull. Eksptl.
    Biol. i Med., No. 8 (1957).
Krushinskii, L. V., Fless, D. A., and Dubrovinskaya, N. V., Peculiarities of the develop-
    ment of parabiotic stages in pathological conditions of the brain caused by the action
    of sound stimulation. Abstracts of Proceedings of an Extramural Scientific Conference
    in Vologda in Memory of N. E. Vvedenskii, 1957.
Krushinskii, L. V., Korzhov, V. A., and Molodkina, L. N., The influence of electric shock
    on pathological conditions caused in rats by sound stimulation. Zhur. Vysshei Nerv.
    Deyatel., Vol. 8, No. 1 (1958).
Krushinskii, L. V. and Fless, D. A., The restorative role of limiting inhibition in the
    pathology of nervous activity. Transactions of the All-Union Society of Physiologists,
    Biochemists, and Pharmacologists, 1958, Vol. 4.
Krushinskii, L. V., Molodkina, L. N. and Levitina, N. A., The time and conditions of re-
    covery of the exhausted process of inhibition during the action of sound stimuli. Zhur.
    Vysshei Nerv. Deyatel., Vol. 9, No. 4 (1959).
Kudienko, I. M., Disturbance of the circulation and respiration in anaphylactic shock in
    animals after extirpation of the thyroid gland in acute experiments. Collected Abstracts
    of Scientific Work of the Faculty of Medicine, Uzhgorod University, L'vov, 1957.
Kupalov, P. S., Experimental neuroses. In: Problems of Corticovisceral Pathology. Izd.
    AMN SSSR, Moscow, 1949.
Kupalov, P. S., Speech of June 30 at a Scientific Session on the Problem of the Physiolog-
    ical Teaching of Academician I. P. Pavlov. Izd. AN SSSR, 1950.
Kupalov, P. S., The physiological organization of the processes of excitation and inhibition
    in the cerebral cortex during conditioned-reflex activity. Zhur. Vysshei Nerv. Deyatel.,
    Vol. 5, No. 4 (1955).
Kupalov, P. S., Some new data concerning the physiology of higher nervous activity. Med.
    Rabotnik, No. 18 (1662) (1958).
Larionov, V. F. and Berdyshev, A. G., Natural molting and its connection with ovulation in
    fowls. Trudy Nauch.-Issled. Inst. Ptitsepromyshlennosti, Vol. 1 (1933).
Liberfarb, A. S., Experiments to study the movement of the gizzard in thyroidectomized
    birds. Zhur. Eksptl. Biol. i Med., Vol. 10, No. 26 (1928).
Liberfarb, A. S., Experiments to show the comparative influence of thyroxin and feeding
    with meat on the movement of the empty gizzard in fowls. Zhur. Eksptl. Biol. i Med.,
    Vol. 11, No. 29 (1928a).
Lindberg, A. A., The action of caffeine on the activity of the cerebral cortex. Doklady
    Akad. Nauk SSSR, Vol. 1, No. 4 (1935).
Lodygina-Kots, N. N., Investigation of the Recognitive Ability of the Chimpanzee. GIZ,
    Moscow, 1923.
Lodygina-Kots, N. N., The Adaptive Motor Habits of the Macaco in Experimental Condi-
    tions. Moscow, 1928.
Lodygina-Kots, N. N., Play, Instincts, Emotions, Habits and Expressive Movements of
    Human and Chimpanzee Children. Published by the State Darwin Museum, Moscow,
    1935.
Lodygina-Kots, N. N., Peculiarities of the intellect of anthropoids as shown by their con-
    structive and tool-handling activity. Reports of a Conference on Psychology. Izd. AN
    SSSR, Moscow, 1957.
Lodygina-Kots, N. N., The Development of the Mind in the Process of Evolution. Izd. Sov.

Nauka, Moscow, 1958.
Lukina, E. V., Vocal reactions of birds of the crow family. Priroda, No. 4 (1957).
Maiorov, F. P. and Troshikhin, V. A., Standard tests for the type of nervous system. In:
     Technique of Study of Conditioned Reflexes, by N. A. Podkopaev, Izd. AN SSSR, 1952.
Malinovskii, A. A., Human Constitution as the Consequence of the Correlation of Signs.
     Candidate dissertation, Moscow, 1935.
Malinovskii, A. A., Physiological correlations in the structure of man. Zhur. Obshchei
     Biol., Vol. 6, No. 4 (1945).
Malinovskii, A. A., Investigation of the processes of excitation and inhibition in 'reflex
     epilepsy' in rats. I. Principal forms of the reaction and their relationship to the
     strength of stimulation. Byull. Eksptl. Biol. i Med., No. 1 (1954).
Mazing, R. A., Variation and inheritance of photoreactions in Drosophila melanogaster.
     Zhur. Óbshchei Biol., Vol. 4, No. 4 (1943).
Mazing, R. A., Choice of medium for laying ova of Drosophila melanogaster. Doklady Akad.
     Nauk SSSR, Vol. 47, No. 4 (1945).
Mazing, R. A., Experiments on selection of Drosophila melanogaster by the ovulating abil-
     ity of the female. Doklady Akad. Nauk SSSR, Vol. 51, No. 5 (1946).
Mazing, R. A., Inheritance of selective ovulating ability in Drosophila melanogaster. Dok-
     lady Akad. Nauk SSSR, Vol. 51, No. 7 (1946a).
Milovanov, V. K. and Smirnov-Ugryumov, P. V., The rational use of pedigree stock in the
     light of Pavlov's teaching. Vestnik Sel'skokhoz. Nauk. Zhivotnovodstvo, No. 5 (1940).
Mogil'nitskii, B. N., Problems of Capillary Permeability in Pathology. Izd. AMN SSSR,
     Moscow, 1949, Vol. 1.
Molodkina, L. N., Physiological Analysis of an Experimental Motor Neurosis Obtained by
     the Method of Sound Stimulation. Candidate dissertation, Moscow, 1956.
Morgan, L., Habit and Instinct. St. Petersburg, 1899.
Napalkov, A. V., Chains of motor conditioned reflexes in pigeons. Abstracts of Proceed-
     ings of the Seventeenth Conference on Problems of Higher Nervous Activity, 1956.
Napalkov, A. V., The conditioned reflex and complex forms of animal behavior. Priroda,
     No. 1 (1958).
Napalkov, A. V., The physiological mechanisms underlying the formation of chains of motor
     conditioned reflexes. Nauch. Dokl. Vysshei Shkoly, No. 2 (1958a).
Napalkov, A. V., The study of the pattern of development of complex systems of conditioned
     reflexes. Vestnik MGU, No. 2 (1958b).
Nikiforovskii, P. M., The Pharmacology of Conditioned Reflexes as a Method of Their
     Study. Dissertation. St. Petersburg, 1910.
Nikiforovskii, P. M., The influence of nerve drugs on conditioned reflexes. Trudy Russk.
     Vrachei, Vol. 78, January–May, 1911.
Oleneva, G. N., Morphological changes in the brain during the action of a sound stimulus.
     Zhur. Nevropat. i Psikhiat., Vol. 55, No. 9 (1955).
Oleneva, G. N., Morphological changes in the brain during the action of sound. Abstracts
     of Proceedings of a Conference of Junior Scientists of the Proletarskii District of
     Moscow (Medical Section), 1958.
Orbeli, L. A., Lectures on Problems of Higher Nervous Activity. Izd. AN SSSR, Moscow–
     Leningrad, 1945.
Orbeli, L. A., Fundamental problems and methods of evolutionary physiology. In: Evolu-
     tion of the Functions of the Nervous System. Medgiz, Moscow, 1958.
Pavlov, I. P., Lectures on the Work of the Cerebral Hemispheres. OGIZ, Moscow–Lenin-
     grad, 1927.
Pavlov, I. P., Twenty Years of Experience of the Objective Study of the Higher Nervous
     Activity (Behavior) of Animals. Biomedgiz, Moscow–Leningrad, 1938.
Pavlov, I. P., The conditioned reflex. In: Great Soviet Encyclopedia, OGIZ RSFSR, 1936.
Pavlov, I. P., Pavlov's Wednesdays. Izd. AN SSSR, Moscow–Leningrad, 1949, Vols. 1, 2,
     and 3.
Pavlov, I. P. and Petrova, M. K., Analysis of some complex reflexes in dogs. The relative
     strength of the centers and their charge. Collection in Honor of K. A. Timiryazev,
     1916.
Pavlova, A. M., The action of caffeine and bromide on the senile central nervous system
     of the dog. Trudy Fiziol. Lab. Akad. I. P. Pavlova, Vol. 8, (1938).
Pavlova, V. I., The formation of a stable, unusual reaction in a dog of a weak type. Trudy
     Fiziol. Lab. im. I. P. Pavlova, Vol. 15 (1949).
Petrov, I. R., The influence of the electric current on animals fed with thyroidin. In: Elec-
     trotrauma. Leningrad, 1939.
Petrov, I. R., Shock and Collapse. Leningrad, 1947.

Petrov, I. R., Etiology, pathogenesis, prevention and treatment of burn shock. Vestnik Khir. im. Grekova, Vol. 77, No. 2 (1956).

Petrova, M. K., Recent Data on the Mechanism of Action of Bromides on the Higher Nervous Activity and on Their Therapeutic Use, Based on Experimental Findings. Izd. VIEM, Moscow, 1935.

Petrova, M. K., The influence of castration on the conditioned-reflex activity and the general behavior of dogs of different types of nervous system. Trudy Fiziol. Lab. Akad. I. P. Pavlova, Vol. 6, No. 1 (1936).

Petrova, M. K., The formation of conditioned nervous connections in an animal soon after castration (a puppy). Trudy Fiziol. Lab. Akad. I. P. Pavlova, Vol. 7 (1937).

Petrova, M. K., The pathophysiology of the higher nervous activity of animals and its relationship to clinical findings. Arkh. Biol. Nauk SSSR, Vol. 46, No. 2 (1937a).

Petrova, M. K., Experimental neuroses. Uspekhi Sovremennoi Biol., Vol. 2, No. 3 (1939).

Petrova, M. K., Changes in the conditioned-reflex activity and general behavior of dogs of different nervous types during prolonged administration of thyroidin. Trudy Fiziol. Lab. im. I. P. Pavlova, Vol. 12 (1945).

Petrova, M. K., Dynamics of nervous processes during inadequate feeding of dog of different nervous types. Izvest. Akad. Nauk SSSR, Ser. Biol., No. 1 (1946).

Platonov, K. I., The study of the importance of hypnotic sleep inhibition as a method of treatment of certain pathological conditions in man. Zhur. Vysshei Nerv. Deyatel., Vol. 2, No. 3 (1952).

Podkopaev, N. A., A special case of motor reaction in a dog in association with the development of inhibition in the cerebral cortex. Trudy Fiziol. Lab. Akad. I. P. Pavlova, Vol. 1, No. 2-3 (1926).

Podkopaev, N. A., The chronic application of external inhibition. Trudy Fiziol. Lab. im I. P. Pavlova, Vol. 12, No. 2 (1945).

Polushkina, S. S., The influence of preliminary parathyroidectomy and castration on the development of traumatic shock in animals. In: Southern RSFSR Branch of the All-Union Society of Physiologists, Biochemists, and Pharmacologists. Abstracts of Proceedings of the Tenth Conference to Commemorate the 15th Anniversary of the Death of Academician I. P. Pavlov. Rostov-on-Don, 1951.

Polyakov, G. I., Relationships between the principal types of neurons in the human cerebral cortex. Zhur. Vysshei Nerv. Deyatel., Vol. 5, No. 3 (1956).

Polyakov, G. I., The increasing complexity of the structure of the neurons of the central nervous system in man, the primates and other mammals. Sovet. Antropol., No. 3 (1958).

Polyakov, G. I., The increasing complexity of the neuronal structure of the brain in man, monkeys and other mammals. Sovet. Antropol., No. 4 (1958).

Polyakov, G. I., The place of the reticular formation in the analyzer systems. Uspekhi Sovremennoi Biol., Vol. 48, No. 2(5) (1959).

Prokopets, I. M., An experimental cataleptoid state. Patol. Fiziol. i Eksptl. Terapiya, Vol. 2, No. 4 (1958).

Prokopets, I. M., An experimental investigation of the protective and restorative role of the functional cataleptoid state. Nauch. Dokl. Vyssh. Shkoly, Biol. Nauki, No. 3 (1958a).

Prokopets, I. M., A physiological analysis of an experimental cataleptoid state, produced by the method of sound stimulation. I. The influence of the character of the excitation, of parathyroidectomy and the drugs (caffeine, bromides and promedol) on the duration of the cataleptoid state. Uch. Zap. Tiraspol. Med. Inst., Vol. 7 (1958b).

Prokopets, I. M., A physiological analysis of an experimental cataleptoid state, produced by the method of sound stimulation. II. The influence of inertia of the process of stimulation on the duration of the cataleptoid state. Uch. Zap. Tiraspol. Med. Inst., Vol. 8 (1958c).

Promptov, A. N., Birds in Nature. Gos. Uchpedgiz, Moscow, 1937.

Promptov, A. N., Elements in the study of the ecological plasticity of some species of birds. Zool. Zhur., Vol. 17, No. 3 (1938).

Promptov, A. N., The species stereotype of behavior and its formation in wild birds. Doklady Akad. Nauk SSSR 27, No. 2 (1940).

Promptov, A. N. and Lukina, E. V., The conditioned-reflex differentiation of desires in birds of the crow family and its biological importance. Doklady Akad. Nauk SSSR, Vol. 46, No. 9 (1945).

Promptov, A. N., Conditioned-reflex components in the instinctive activity of birds. Fiziol. Zhur. SSSR, Vol. 32, No. 1 (1946).

Promptov, A. N., Evolutionary and biological peculiarities of the orienting reaction in certain ecologically specialized species of birds. Trudy Inst. Evol. Fiziol. i Patol.

Vysshei Nerv. Deyatel. im I. P. Pavlova, Vol. 1 (1947).

Promptov, A. N., The physiological mechanism and biological factors of the formation of biocomplexes of feeding activity in birds. Proceedings of the Thirtieth Conference on Physiological Problems in Memory of I. P. Pavlov. 1948.

Promptov, A. N., Essays on the Biological Adaptation of the Behavior of Birds of the Crow Family. Izd. AN SSSR, Moscow–Leningrad, 1956.

Pugachev, A. G., Changes in the conditioned-reflex activity of birds (pigeons) with different types of nervous system under the influence of experimental hyperthyroidism. Uch. Zap. Tomsk. Pedagog. Inst., Vol. 10, (1953).

Pugachev, A. G., The influence of experimental hyperthyroidism on nervous crises in birds (pigeons). Uch. Zap. Tomsk. Pedagog. Inst., Vol. 11 (1954).

Pyshkina, S. P., Changes in conditioned-reflex reactions under the influence of strychnine. Arkh. Biol. Nauk SSSR, Vol. 5, No. 3 (1939).

Rikman, V. V., Disturbance of the normal nervous activity of the dog under the influence of strong, unusual stimuli. Trudy Fiziol. Lab. Akad. I. P. Pavlova, Vol. 3, No. 1 (1928).

Rikman, V. V., The detection of old traces of excitation of the centers of the defensive reaction as the analogoue of a traumatic neurosis. Trudy Fiziol. Lab. Akad. I. P. Pavlova, Vol. 4 Nos. 1 and 2 (1932).

Roginskii, Ya. Ya., Investigation of the association between physique and motor function. Antropol Zhur., No. 3 (1937).

Rokotova, N. A., Chains of motor conditioned reflexes in dogs. Zhur. Vysshei Nerv. Deyatel., Vol. 4, No. 6 (1954).

Rosin, Ya. A., The role of the vegetative nervous system in the genesis of shock. In: Shock (Transactions of a Conference on the Problem of Shock). Kiev, 1938.

Rozental', I. S., Characteristics of the orienting and defensive reflexes. Arkh. Biol. Nauk SSSR, Vol. 30, No. 1 (1930).

Rozental', I. S., Training and nervous type. Arkh. Biol. Nauk SSSR, Vol. 41, No. 2 (1936).

Rozhanskii, N. A., Relationship between cortical and subcortical inhibition. Abstracts of Proceedings of a Scientific Session to Commemorate the Tenth Anniversary of the Death of I. P. Pavlov, 1946.

Rozhanskii, N. A., Ways of studying conduction in the subcortical region of the brain and the brainstem. Proceedings of the Seventh All-Union Congress of Physiologists, Biochemists, and Pharmacologists. Moscow, 1947.

Sadovnikova-Kol'tsova, M. P., A genetic analysis of the mental ability of rats. Zhur. Eksptl. Biol. i Med., No. 1 (1925).

Sadovnikova-Kol'tsova, M. P., A genetic analysis of the mental ability of rats, Part 2. Zhur. Eksptl. Biol. i Med., Ser. A, Vol. 4, No. 1 (1928).

Sadovnikova-Kol'tsova, M. P., A genetic analysis of the mental ability of rats, Part 3. Zhur. Eksptl. Biol., Vol. 7, No. 3 (1931).

Sadovnikova-Kol'tsova, M. P., A genetic analysis of the mental ability of rats, Part 4. Biol. Zhur., Vol. 2, No. 2–3 (1933).

Sadovnikova-Kol'tsova, M. P., A genetic analysis of the mental ability of rats, Part 5. Biol. Zhur., Vol. 3, No. 4 (1934).

Sadovnikova-Kol'tsova, M. P., Analysis of the temperament of wild rats and their hybrids. Biol. Zhur., Vol. 7, No. 3 (1938).

Sechenov, I. M., The reflexes of the brain. In: Selected Works. Izd. VIEM, Moscow, 1935.

Sechenov, I. M., Who is to develop psychology, and how? In: Selected Works. Izd. VIEM, Moscow, 1935a.

Semiokhina, A. F., An electrophysiological investigation of the auditory and motor analyzers in a model of experimental motor neurosis. Zhur. Vysshei Nerv. Deyatel., Vol. 8, No. 2 (1958).

Serebryannikov, I. S., The central nervous system in the pathogenesis of shock. Program and Abstracts of Proceedings of the Sixteenth Session of the Arkhangelsk Medical Institute, 1951.

Sereiskii, M. Ya., A new method of treatment of epilepsy. Report I. Zhur. Nevropat. i Psikhiat., Vol. 55, No. 9 (1955).

Servit, Z., The development of an experimental epileptic fit at different stages of the phylogenesis of vertebrates. Fiziol. Zhur. SSSR, Vol. 38, No. 6 (1952).

Servit, Z., The relationship between excitation and inhibition in the pathophysiology of the epileptic fit. Zhur. Vysshei Nerv. Deyatel., Vol. 5, No. 4 (1955).

Slonim, A. D., The study of specialized reflex acts in mammals. In: The Evolution of the Functions of the Nervous System. Medgiz, Moscow, 1958.

Sokolov, E. N., Perception and the Conditioned Reflex. Izd. MGU, Moscow, 1958.

Speranskaya, E. N., Baranova, F. D., Belovintseva, M. F., Mityushov, M. I., and Tendler,

D. S., Problems of the nervous regulation of the activity of the glands of internal secretion. In: Eighth All-Union Congress of Physiologists, Biochemists, and Pharmacologists, 1955.

Speranskii, A. D., Cowardice and inhibition. Transactions of the Second All-Union Congress of Physiologists, 1926.

Speranskii, A. D., The influence of strong, destructive stimuli on a dog with an inhibited type of nervous system. Trudy Fiziol. Lab. Akad. I. P. Pavlova, Vol. 2, No. 1 (1927).

Steshenko, A. P., The physiological analysis of shock and hemorrhagic states. Nauch. Dokl. Vysshei Shkoly (Biol. Nauki), No. 1 (1959).

Tendler, D. S., The Influence of Methylthiouracil and the Functional Disturbances of the Thyroid Gland Caused Thereby on the Extero- and Interoceptive Conditioned Reflexes. Candidate dissertation, Leningrad, 1952.

Timofeeva, T. A., The course of study of experimental genetics of higher nervous activity. Abstracts of Proceedings of the Ninth Conference on Physiological Problems, Leningrad, 1941.

Tonkikh, A. V., Experimental hyperthyroidism. Report 3. Experimental administration of extract of the anterior lobe of the hypophysis. Fiziol. Zhur. SSSR, Vol. 26, No. 6 (1939).

Tsitovich, I. S., The Genesis and Formation of Natural Conditioned Reflexes. Dissertation, 1911.

Ukhtomskii, A. A., The dominant as a working principle of nerve centers. Russk. Fiziol. Zhur., Vol. 6, Nos. 1, 2, and 3 (1923).

Ukhtomskii, A. A., Collected Works. Leningrad, 1945, Vol. 4.

Ukhtomskii, A. A., The state of excitation in the dominant. In: Collected Works, Leningrad, 1950, Vol. 1.

Usievich, M. A., The functional state of the cerebral cortex and the activity of the internal systems of the body. Advances in Medicine (Higher Nervous Activity), Izd. AMN SSSR, Moscow, 1949, No. 14.

Usievich, M. A., Artem'ev, E. I., Alekseeva, T. G., and Stepanova, A. D., The interrelationship between the activity of the sex and thyroid glands and the higher nervous activity. Fiziol. Zhur. SSSR, Vol. 25, No. 4 (1938).

Vagner, V. A., The biological Basis of Comparative Psychology. St. Petersburg, 1910–1913, Vols. 1 and 2.

Vagner, V. A., The appearance and development of mental abilities. In: From Reflexes to Instincts. Leningrad, 1925, No. 3.

Val'kov, A., Experimental study of the nervous activity of thyroidectomized puppies. In: Collection to Commemorate the Seventieth Birthday of Academician I. P. Pavlov, 1925.

Van Bin', Conditioned Reflexes in Some Functional and Organic Lesions of the Cerebral Cortex. Candidate dissertation, Moscow, 1958.

Vasil'ev, Yu. A., The mechanism of Parfenov's reaction. Russk. Fiziol. Zhur., Vol. 6, No. 4, 5, and 6 (1924).

Vasil'ev, B. A., The Physiological Analysis of Certain Forms of Behavior of Birds. Doctorate dissertation, Leningrad, 1941.

Vasil'eva, V. M., Changes in the Electrical Activity of the cortical division of the motor analyzer in white rats during a reflex epileptiform fit. Byull. Eksptl. Biol. i Med., No. 1 (1957).

Vetyukov, I. A., The problem of experimental neurosis in dogs. Trudy Fiziol. Nauchno-Issled. Inst. LGU, No. 17 (1936).

Volokhov, A. A., Laws of Ontegenesis of Nervous Activity. Izd. AN SSSR, Moscow–Leningrad, 1951.

Voronin, L. G., Against the anti-Pavlovian concepts of Academician Beritashvili. Fiziol. Zhur. SSSR, Vol. 37, No. 3 (1951).

Voronin, L. G., Analysis and Synthesis of Complex Stimuli in Higher Animals. Medgiz, Leningrad, 1952.

Voronin, L. G., The Conditioned Reflex, a Universal Adaptive Phenomenon in the Animal World. Izd. Znanie, 1954.

Voronin, L. G., Lectures on the Comparative Physiology of Higher Nervous Activity. Izd. MGU, 1957.

Voronin, L. G., General and specific factors in the phylogenesis of higher nervous activity. In: Evolution of the Functions of the Nervous System, edited by D. A. Biryukov. Medgiz, Leningrad, 1958.

Vvedenskii, N. E., The relationship between stimulation and excitation in tetanus. In: Collected Works. 1934, Vol. 2.

Vvedenskii, N. E., Excitation, inhibition and narcosis. In: Collected Works. 1935, Vol. 4.

Vyrzhikovskii, S. N., The inhibited, weak type of nervous system. Trudy Fiziol. Lab. Akad. I. P. Pavlova, Vol. 3, No. 1 (1928).

Vyrzhikovskii, S. N., The so-called primary reflex of biological caution. Conference on Problems of Higher Nervous Activity on the First Anniversary of the Death of Academician I. P. Pavlov, 1937.

Vyrzhikovskii, S. N. and Maiorov, F. P., The influence of training on the pattern of higher nervous activity in dogs. Trudy Fiziol. Lab. Akad. I. P. Pavlova, Vol. 5 (1933).

Yakovleva, V. V., The physiological mechanism of formation of difficult differentiation. Trudy Fiziol. Lab. im. I. P. Pavlova, Vol. 9, (1940).

Yakovleva, V. V., The formation of a pathological passive-defensive reflex in a dog of strong type. Trudy Fiziol. Lab. im. I. P. Pavlova, Vol. 15 (1949).

Zavadovskii, M. M., Sex and the development of its characteristics. In: The Analysis of Morphogenesis. Gosizdat, Moscow, 1922.

Zavadovskii, B. M. and Rokhlina, M. L., Conditioned reflexes in normal and hyperthyroidized fowls. Med. Biol. Zhur., No. 3 (1927).

Zavadovskii, B. M., Azimov, G. I., and Zakharov, V. P., The influence of the thyroid gland on the higher nervous activity of dogs. Med. Biol. Zhur., No. 1 (1929).

Zavadovskii, B. M., Zakharov, V. P., and Zlotov, M. O., The influence of the thyroid gland on the higher nervous activity of dogs. Med. Biol. Zhur., No. 3 (1929).

Zavadovskii, B. M. and Zlotov, M. O., The influence of the thyroid gland on the higher nervous activity of dogs, Report 4. Med. Biol. Zhur., No. 4 (1929).

Zeval'd, L. O., The influence of training conditions on the pattern of higher nervous activity in dogs. Trudy Fiziol. Lab. Akad. I. P. Pavlova, Vol. 8, (1938).

Zeval'd, L. O., The effect of caffeine and its combination with bromide on higher nervous activity. Trudy Fiziol. Lab. Akad. I. P. Pavlova, Vol. 8 (1938a).

Zeval'd, L. O., The influence of extirpation of the parathyroid glands on the conditioned-reflex activity of the dog. Trudy Inst. Evol. Fiziol. i Patol. Vysshei Nerv. Deyatel. im I. P. Pavlova, Vol. 1 (1947).

Zhuravlev, I. N., The influence of strychnine, cocaine and nicotine on conditioned and unconditioned food salivatory reflexes. Abstracts of Proceedings of the Third Conference on Physiological Problems, 1938.

Zimkin, N. V., Disturbance of the normal balance in the cerebral cortex and its restoration by the influence of caffeine and differentiation. Trudy Fiziol. Lab. Akad. I. P. Pavlova, Vol. 3, No. 1 (1928).

NON-SOVIET LITERATURE

Altmann, M., Behavior of the sow in relation to the sex cycle. Proceedings of the American Physiology Society. Am. J. Physiol., Vol. 126, No. 3 (1939).

Anderson, E. E., The sex hormones and emotional behavior. I. The effect of sexual receptivity upon timidity in the female rats. J. Genet. Psychol., Vol. 56 (1940a).

Anderson, E. E., The sex hormones and emotional behavior. III. The effect of castration upon timidity in male and female rats. J. Genet. Psychol., Vol. 56 (1940).

Anderson, E. E. and Anderson, S. F., The sex hormones and emotional behavior. II. The influence of the female sex hormones upon timidity in normal and castrated female rats. J. Genet. Psychol., Vol. 56 (1940).

Anderson, O. D., The spontaneous neuromuscular activity of various pure breeds of dogs and of interbreed hybrids of the first and second generation. Am. J. Physiol., Vol. 126, No. 3 (1939).

Anderson, O. D., The role of the glands of internal secretion in the production of behavioral types in the dog. In: C. Stockard, The Genetic and Endocrinic Basis for Differences in Form and Behavior. 1941.

Armstrong, E. A., The nature and function of displacement activities. In: Physiological Mechanisms in Animal Behavior. Symposium of the Society for Experimental Biology, No. IV,. Academic Press, New York, 1950.

Ashby, W. R., Design for an intelligence-amplifier. In: C. E. Shanon and J. MacCarthy, ed. Automata Studies (Annals of Mathematical Studies No. 34), Princeton University Press, Princeton, 1956.

Baege, B., Zur Entwicklung der Verhaltensweisen junger Hunde in den ersten drei Lebensmonaten. Zeit. f. Hundeforsch., Vol. 3, Nos. 1 and 2 (1933).

Bagg, H., Individual differences and family resemblances in animal behavior. Am. Natural., Vol. 50, No. 592 (1916).

Bibliography

257

Bauer, F., Genetic and experimental factors affecting social reactions in male mice. J. Comp. and Physiol. Psychol., Vol. 49, No. 4 (1956).

Beach, F. A., The neural basis of innate behavior. I. Effect of cortical lesions upon the maternal behavior pattern in the rat. J. Comp. Psychol., Vol. 24, No. 3 (1937).

Beach, F. A., The neural basis of innate behavior. III. Comparison of learning ability and instinctive behavior in the rat. J. Comp. Psychol., Vol. 28, No. 2 (1939).

Beach, F. A., Hormones and Behavior. New York, 1948.

Beach, F. A. and Weaver, T. H., Noise-induced seizures in the rat and their modification by cerberal injury. J. Comp. Neurol., Vol. 79., No. 3 (1943).

Beach, F. A. and Jordan, L., Effects of sexual reinforcement upon the performance of male rats in a straight runway. J. Comp. and Physiol. Psychol., Vol. 49, No. 2 (1956).

Beeman, E. A., Male hormone as a cause of aggressive behavior in male mice. Anatom. Rec., Vol. 96, No. 4 (1946).

Bella-Bella, D., Characteres et nature de modifications de la pression arterielle au cours de electrochoc. Arch. intern. physiol., Vol. 62, No. 2 (1954).

Bierens de Haan, J. A., Neuere Untersuchungen ueber die hoeheren Formen der tierischen Intelligenz. Verhandl. Dtsch. Zool. Ges., 1931a.

Bierens de Haan, J. A., Werkzeuggebrauch und Werkzeugherstellung bei einem niederen Affen. Z. vergleich. Physiol., Vol. 13 (1931b).

Bitterman, M. E., Behavior disorder as a function of the relative strength of antagonistic response-tendencies. Psychol. Rev., Vol. 51, No. 6 (1944).

Boelter, M. and Greenberg, D. M., Severe calcium deficiency in growing rats. I. Symptoms and pathology. J. Nutrition, Vol. 21, No. 1 (1941).

Boll, L., The female sex cycle as a factor in learning in the rat. Am. J. Physiol., Vol. 78 (1926).

Bonvallet, M., Dell, P., and Hiebel, G., Tonus sympatique et activité électrique corticale. Electroceph. clin. neurophisiol., Vol. 6, p. 137 (1954).

Brody, E., Genetic basis of spontaneous activity in the albino rat. Biol. Abstr., Vol. 17, No. 1 (1943).

Brody, E., A note of the genetic basis of spontaneous activity in the albino rat. J. Comp. and Physiol. Psychol., Vol. 43, No. 4 (1950).

Brull, H., Das Leben Deutscher Greifvoegel. Fischer Verlag, Jena, 1937.

Bugbee, E. P. and Simond, A. E., The increase of voluntary activity of ovariectomized albino rats caused by injections of ovarian follicular hormone. Endocrinology. Vol. 10 (1926).

Bures, J., Susceptibility to convulsions in reflex epilepsy in the ontogenesis of rats and mice. Czechoslov. Physiol., Vol. 2, No. 3 (1953).

Bures, J., Experiments on the electrophysiological analysis of the generalization of an epileptic fit. Czechoslov. Physiol., Vol. 2, No. 4 (1953a).

Busnel, R. G., Lehman, A., and Busnel, M. C., Étude de la crise audiogène de la souris comme test psychopharmacologique: son application aux substances de type tranquilliseur. Pathol. Biologie, Vol. 34, No. 9-10 (1958).

Buytendijk, F., The Mind of the Dog. London, 1935.

Campbell, B. A., and Sheffield, F. D., Relations of random activity to food deprivation. J. Comp. Physiol. Psychol., Vol. 46 (1953).

Carpenter, C. R., Sexual behavior of free ranging Rhesus monkey (Macaca mulatta). II. Periodicity of oestrus homosexual, autoerotic and noncorformist behavior. J. Comp. Psychol., Vol. 33, No. 2 (1942).

Ceni, C., Die endocrinen Faktoren der Mutterliebe und die psychische Feminierung von Maennchen. Schweiz. Archiv fuer Neurol. und Psychiatrie, Vol. 21, No. 1 (1927).

Ceni, C., Ueber die Verwandlung des Geschlechtstriebes in den Muttertrieb beim Weibchen und beim Maennchen. Zeitschr. Sexualwissenschaft und Sexualpolitik, Vol. 26, No. 1 (1929).

Chambers, R. and Zweifach, B. W., Intercellular cement and capillary permeability. Physiol. Rev., Vol. 27, No. 3 (1947).

Coburn, C., Heredity of Wildness and Savageness in Mice. 1922, Vol. 4, No. 5.

Commins, W. D., The effects of castration at various ages upon the learning ability of male albino rats. J. Comp. Psychol., Vol. 14, No. 1 (1932).

Corey, S. M., Sex differences in maze learning by white rats. J. Comp. Psychol., Vol. 10 (1930).

Crisler, G., Booher, W., Van Liere, E., and Hall. The effect of feeding thyroid on the salivary condition reflex induced by morphin. Am. J. Physiol., Vol. 103, No. 4 (1933).

Dawson, W., Inheritance of wildness and tameness in mice. Genetics, Vol. 17, No. 3 (1932).

Dawson, W. and Katz, R., Preliminary report on variation in ability of dogs to master a multiple choice situation. Genetics, Vol. 25, No. 1 (1940).

Donaldson, H. H., The Rat. Philadelphia, 1924.

Durrant, E. P., Studies on vigor. I. Effect of adrenal extirpation on activity of the albino rat. Am. J. Physiol., Vol. 70, No. 1 (1924).

Eayrs, J., Spontaneous activity in the rat. Brit. J. Animal Behavior, Vol. 46 (1954).

Eccles, J., The neurophysiological basis of mind. In: The Principles of Neurophysiology. Oxford, 1953.

Eccles, J., The Physiology of Nerve Cells. Oxford, 1957.

Erhardt, K., Beitraege zur Hypophysenforderlapenreaktion unter besonderer Beruecksichtigung der Ascheim Zondekschen Schwangerschaftsreaktion. Klin. Wochenschr., No. 44 (1929).

Farris, E. J. and Jeakel, E. H., Sex and increasing age as factors in the frequency of audiogenic seizures in albino rats. J. Comp. Psychol., Vol. 33, No. 4 (1942a).

Farris, E. J. and Jeakel, E. H., The effect of age upon susceptibility to audiogenic seizures in albino rats. J. Comp. Psychol., Vol. 33, No. 2 (1942b).

Finger, F. W., Abnormal animal behavior and conflict. Psychol. Rev., Vol. 52, No. 4 (1945).

Finger, F. W., Convulsive behavior in the rat. Psychol. Bull., Vol. 44, No. 3 (1947).

Fischel, W., Intelligenz und Einsicht der Affen. Arch. Neerland. de Zoology, Vol. 10, Suppl. 2 (1953).

Fischel, W., Leben und Erlebnis bei Tieren und Menschen. Munich, 1949.

Fischel, W., Die hoeheren Leistungen der Wirbeltiergehirne. Leipzig, 1956.

Frings, M. and Frings, H., Audiogenic seizures in the laboratory mouse. J. Mammalogy, Vol. 33, No. 1 (1952).

Gans, H. M., Studies on vigor. XIII. Effect of early castration on the voluntary activity of male albino rats. Endocrinology, Vol. 11, No. 2 (1927).

Gentry, E. and Dunlap, K., An attempt to produce neurotic behavior in rats. J. Comp. Psychol., Vol. 33, No. 1 (1942).

Gould, L. and Morgan, C. T., Hearing in the rat at high frequencies. Science, Vol. 94 (1941).

Gould, L. and Morgan, C. T., Patterns of electrogenic seizures in rats: their relation to stimulus intensity and to audiogenic seizures. J. Comp. Psychol., Vol. 38, No. 4 (1945).

Greenberg, D. M., Boelter, M. D., and Knopf, B. W., New observations on the effect of calcium deprivation. Science, Vol. 89, No. 2297 (1941).

Greenberg, D. M., Boelter, M. D., and Knopf, B. W., Factors concerned in the development of tetany by the rat. Am. J. Physiol., Vol. 137, No. 2 (1942).

Greenberg, D. M. and Tufts, E. V., The nature of magnesium tetany. Am. J. Physiol., Vol. 121, No. 2 (1932).

Griffiths, W. J., Transmission of convulsions in the white rat. J. Comp. Psychol. Vol. 34, No. 2 (1942).

Griffiths, W. J., The effect of thiamine hydrochloride on the incidence of audiogenic seizures among selectively bred albino rats. J. Comp. Psychol., Vol. 38, No. 2 (1945).

Heinroth, O., Reflektorische Bewegungen bei Voegeln. J. Ornithologie, Vol. 66, Nos. 1 and 2 (1918).

Heinroth, O., Aus dem Leben der Voegel. Berlin, 1938.

Heller, R., Spontaneous activity in male rats in relation to tetis hormone. Endocrinology, Vol. 16, No. 6 (1932).

Henderson, I., Haggard, H. W., and Coburn, R. C., The therapeutic use of carbon dioxide after anesthesia and operation. J. Am. Med. Association, Vol. 20 (1920).

Hinde, R. A., Alternative motor patterns in chaffinch song. In: Animal Behavior, 1958, Vol. 6, No. 3-4.

Hitchcock, F. A., Studies on vigor. V. The comparative activity of male and female albino rats. Am. J. Physiol., Vol. 75, No. 1 (1925).

Hoskins, R. G., Endocrine factors influencing bodily vigor. Am. J. Med. Ass., Vol. 85 (1925).

Hoskins, R. G., Studies on vigor. II. The effect of castration on voluntary activity. Am. J. Physiol., Vol. 75, No. 2 (1925a).

Humphrey, E. and Warner, L., Working Dogs. Baltimore, 1934.

James, W. T., Morphological form and its relation to behavior, Section IV. In: C. Stockard, The Genetic and Endocrine Basis for Differences in Form and Behavior, 1941.

Jasper, H. H., Diffuse projection systems: the integrative action of the thalamic reticular system. Electroencephalog. and Clin. Neurophysiol., Vol. 1 (1949).

Keeler, C. E. and Trimble, H. C., Inheritance of position preference in coach dogs. J.

Heredity, Vol. 31, No. 2 (1940).
Kleitmann, N. and Titelbaum, S., Effect of thyroid administration upon differentiating ability of dogs. Am. J. Physiol., Vol. 115, No. 1 (1936).
Konorski, J., Mechanism of learning. In: Physiological Mechanisms in Animal Behavior. Symposium of the Society for Experimental Biology, No. IV. Academic Press, New York, 1950.
Kosman, M. E. and Damour, F. E., The effect of hypoxia, hyperoxia, and hypercapna upon audiogenic seizures. J. Comp. and Physiol. Psychol., Vol. 49, No. 2 (1956).
Kruse, H. D., Orent, E. R., and McCollum, E. V., Studies on magnesium deficiencies in animals. Symptomatology resulting from magnesium deprivation. J. Biol. Chem., Vol. 98, No. 2 (1932).
Kunde, M. and Neville, M., Changes in reflex response and electrical excitation of peripheral motor nerves in experimental hypo- and hyperthyroidism. Am. J. Physiol. Vol. 92, No. 2 (1930).
Kuo, Z. Y., How are instincts acquired. Psychol. Rev., Vol. 29, No. 5 (1922).
Kuo, Z. Y., Ontogeny of embryonic behavior in aves. V. Psychol. Rev., Vol. 19 (1932).
Kuo, Z. Y., Ontogeny of embryonic behavior in aves. III. The structural and environmental factors in embryonic behavior. J. Comp. Psychol., Vol. 13, No. 2 (1932a).
Kuo, Z. Y., Ontogeny of embryonic behavior in aves. J. Exptl. Zool., Vol. 61, No. 3 (1932b).
Kuo, Z. Y., Ontogeny of embryonic behavior in aves. IV. J. Comp. Psychol., Vol. 19, No. 1 (1932c).
Kuo, Z. Y. and Shen, T. Ontogeny of embryonic behavior in aves. X. Gastric movements of the chick embryo. J. Comp. Psychol., Vol. 21, No. 1 (1936).
Larionow, W., Zur Frage der Bedeutung des Schilddruesenhormons bei Federwechselprozess. Biol. Gener., Vol. 11, No. 1 (1936).
Lashely, K. S., Experimental analysis of instinctive behavior. Psychol. Rev., Vol. 45, No. 6 (1938).
Lee, M. O. and Buskirk, E. F., Studies on vigor. XV. The effect of thyroidectomy on spontaneous activity in the rat with a consideration of the relation of the basal metabolism to spontaneous activity. Am. J. Physiol., Vol. 84, No. 2 (1928).
Lehmann, A., and Busnel, R., Etude sur la crise audiogène. Arch. Scien. Physiol., Vol. 13, No. 2 (1959).
Leopold, A., The nature of heritable wildness in turkeys. The Condor, Vol. 46, No. 4 (1944).
Liddell, H., The effect of thyroidectomy and some unconditioned responses of the sheep and goat. Am. J. Physiol., Vol. 75 (1925).
Lorentz de Nō, R., Cerebral cortex. In: J. F. Fulton, Physiology of the Nervous System. Oxford University Press, 1951.
Lorenz, K., Der Kumpan in der Umwelt des Vogels. J. Ornithologie, Vol. 85, Nos. 2 and 3 (1935).
Lorenz, K., Ueber die Bildung des Instinktbegriffes. Die Naturwissenschaften, Vol. 25, Nos. 19, 20, and 21 (1937).
Lorenz, K., Ueber den Begriff der Instinkhandlung. Instinctus. Folia Biotheor., Ser. B., Vol. 11 (1937a).
Lorenz, K., Vergleichende Verhaltensforschung. Zool. Anzeiger, Suppl. 12. Verhandlungen d. deutschen zool. Gesel. 41. Vol. 1-5 (1939).
Lorenz, K., The comparative method in studying innate behavior patterns, In: Physiological Mechanisms in Animal Behavior. Symposium of the Society for Experimental Biology, No. IV. Academic Press, New York, 1950.
Lorenz, K., The objectivistic theory of instinct. In: L'instinct dans le comportement des animaux et de l'homme. Paris, 1956a.
Lorenz, K., Plays and vacuum activities. In: L'instinct dans le comportement des animaux et de l'homme. Paris, 1956b.
Maier, N. R. F. and Glaser, N. M., Studies of abnormal behavior in the rat. V. The inheritance of the neurotic pattern. J. Comp. Psychol., Vol. 30, No. 2 (1940).
Maier, N. R. F. and Glaser, N. M., Studies of abnormal behavior in the rat. X. The influences of age and sex on the susceptibility to seizures during auditory stimulation. J. Comp. Psychol., Vol. 34, No. 1 (1942).
Maier, N., Studies of abnormal behavior in the rat. XIX. Strain differences in the inheritance of susceptibility to convulsions. J. Comp. Psychol., Vol. 35, No. 3 (1943).
Marchlewski, T., Genetic studies on domestic dogs. International Bulletin Academy of Sciences, Cracow, Vol. 2, pp. 117-145 (1930).
McNemar, Q. and Stone, C., Sex differences in rats on three learning tasks. J. Comp. Psychol., Vol. 14, No. 1 (1932).

Menzel, K., Welpe und Umwelt. Zeitschr. Hundeforsch., Neue Folge, Vol. 3 (1937).

Morgan, C. T., The latency of audigenic seizures. J. Comp. Psychol., Vol. 32, No. 2 (1941).

Morrell, F., Lasting changes in synoptic organization produced by continuous neuronal bombardment. In: Background Material for Pavlov Conference. New York, October 1960.

Moruzzi, G. and Magoun, H., Brain stem reticular formation and activation of the EEG. Electroencephalog. and Clin. Neurophysiol., Vol. 1 (1949).

Patton, R. A., Karn, H. W., and King, C. G., Studies on the nutritional basis of abnormal behavior in albino rats. II. Further analysis of the effects of inanition and vitamin B_1 on convulsive seizures. J. Comp. Psychol., Vol. 33, No. 2 (1942).

Patton, R. A., Karn, H. W., and King, C. G., Studies on the nutritional basis of abnormal behavior in albino rats. III. The effects of different levels of vitamin B_1 intake on convulsive seizures; the effect of other vitamins in the B-complex and mineral supplement on convulsive seizures. J. Comp. Psychol., Vol. 34, No. 1 (1942a).

Pawlowski, A. and Scott, J. P., Hereditary differences in the development of dominance in litters of puppies. J. Comp. and Physiol. Psychol., Vol. 49, No. 4 (1956).

Phillips, J., Note on wildness in ducklings. J. Anim. Behav., Vol. 2 (1912).

Richter, C. P., The effects of early gonadectomy on the gross body activity of rats. Endocrinology, Vol. 17, No. 4 (1933).

Richter, C. P. and Wislocki, G. B., Activity studies on castrated male and female rats with testicular grafts in correlation with histological studies of the grafts. Am. J. Physiol., Vol. 86 (1928).

Riddle, O., Factors in the development of sex and secondary sexual characteristics. Psychol. Rev., Vol. 38, No. 1 (1931).

Riddle, O. Aspects and implications of the hormonal control of the maternal instinct. Proceedings of the American Philosophical Society, 1935, Vol. 75, No. 6.

Rosenblatt, J. S. and Aronson, L. R., The influence of experience on the behavioral effects of androgen in prepuberally castrated male cats. In: Animal Behavior, 1958, Vol. 6, No. 3-4.

Rundquist, E., Inheritance of spontaneous activity in rats. J. Comp. Psychol., Vol. 16 (1933).

Rundquist, E. A. and Heron, W. T., Spontaneous activity and maze learning. J. Comp. Psychol., Vol. 19, No. 2 (1935).

Russel, E. S., Conation and perception in animal learning. Biol. Rev., Vol. 7, (1932).

Russel, E. S., The Behavior of Animals: An Introduction to Its Study. London, 1946.

Scheibel, M. E. and Scheibel, A. B., Structural substrates for integrative patterns in the brain stem reticular core. In: Reticular Formation of the Brain. Henry Ford Hospital Symposium. Boston, 1958.

Schmidt, D., Zur Psychologie hundeartiger Tiere. Natur und Volk, Vol. 70, No. 6 (1940).

Schneirla, T. C., L'instinct (Coll. Int. sur l'instinct animal, 1954), Chapter 10, pp. 387-452, Masson, Paris, 1956.

Scott, J. P., Genetic differences in the social behavior of inbred strain in mice. J. Heredity, Vol. 33, No. 1 (1942).

Scott, J. P., Experimental modification of aggressive and defensive behavior in the C-57 inbred strain in mice. Abstr. Genetics, Vol. 30, No. 1 (1945).

Seward, J. P., Aggressive behavior in the rat. I. General characteristics: age and sex differences. J. Comp. Psychol., Vol. 38, No. 4 (1945).

Shyrley, M., Studies of activity. I. Consistency of the drum method of measuring the activity of the rat. J. Comp. Psychol., Vol. 8, No. 1 (1928a).

Shyrley, M., Studies of activity. II. Activity rhythms, age, and activity: activity after rest. J. Comp. Psychol., Vol. 8 (1928b).

Slonaker, J. R., The normal activity of the albino rat from birth to natural death, its rate and the duration of life. J. Anim. Behav., Vol. 2 (1912).

Slonaker, J. R., Analysis of daily activity of the albino rat. Am. J. Physiol. Vol. 73, No. 2 (1925).

Slonaker, J. R., The effect of the follicular hormone on old albino rats. Am. J. Physiol., Vol. 81, No. 2 (1927).

Slonaker, J. R., The effect of the excision of different sexual organs on the development, growth, and longevity of the albino rat. Am. J. Physiol., Vol. 93, No. 2 (1930).

Smith, K., Quantitative analysis of the pattern of activity in audioepileptic seizures in rats. J. Comp. Psychol., Vol. 32, No. 2 (1941).

Stephanitz, M., Der Deutsche Schaeferhund in Wort und Bild. 1932.

Stockard, C. R. et al., The Genetic and Endocrinic Basis for Differences in Form and

Behavior. Philadelphia, 1941.

Thorne, F.C., The inheritance of shyness in dogs. J. Genetic Psychol., Vol. 65 (1944).

Thorpe, W.H., Learning and Instinct in Animals. London, 1958.

Tinbergen, N., The hierarchial organization of nervous mechanisms underlying instinctive behavior. In: Physiological Mechanisms in Animal Behavior. Symposium of the Society for Experimental Biology, No. IV. Academic Press, New York 1950.

Tinbergen, N., The Study of Instinct. Oxford, 1955.

Tolman, E.C., Inheritance of maze-learning ability in rats. J. Comp. Psychol., Vol. 4, No. 1 (1924).

Trimble, H.C. and Keeler, C.E., Preference of Dalmatian dogs for particular position in coach running and inheritance of their character. Nature, Vol. 7, No. 4 (1939).

Tsai, L.S., Sex glands and adaptive ability. Science, Vol. 71 (1930).

Tuttle, W.W., and Dyskshorn, S., Effect of castration and ovariectomy on spontaneous activity and ability to learn. Proceedings of the Society of Experimental Biology and Medicine, 1928, Vol. 25.

Tyron, R.C., Studies in individual differences in maze ability. II. The determination of individual differences by age, weight, sex, and pigmentation. J. Comp. Psychol., Vol. 12, No. 1 (1931).

Ulrich, J.L., Distribution of Effort in Learning in a White Rat. New York, 1915.

Ulrich, J., The social hierarchy in albino mice. J. Comp. Psychol., Vol. 25, No. 3 (1938).

Utsurikawa, W.A., Temperamental differences between outbred and inbred strains in albino rats. J. Anim. Behav., Vol. 7 (1917).

Uttley, A.M., Temporal and spatial patterns in a conditional probability machine. In: C.E. Shannon and J. MacCarthy, ed., Automata Studies (Annals of Mathematical Studies No. 34), Princeton University Press, Princeton, 1956.

Vicari, E., Mode of inheritance of reaction time and degrees of learning in mice. J. Experimental Zoology, Vol. 54, No. 1 (1929).

Wang, G.H. The relation between spontaneous activity and oestrus cycle in the white rat. In: Comparative Psychology. 1923, Vol. 2, No. 1.

Wang, G.H., Richter, D.P., and Guttmacher, A.F., Activity studies on male castrated rats with ovarian transplants and correlation of the activity with the histology of the graft. Am. J. Physiol., Vol. 73, No. 3 (1925).

Wang, G.H., Stein, P., and Brown, V.W., Effects of transections of central neuraxis on galvanic skin reflex in anesthetized cats. J. Neurophysiol., Vol. 19, p. 340 (1956).

Weiner, H.M. and Morgan, C.T., Effect of cortical lesions upon audiogenic seizures. J. Comp. Psychol., Vol. 38, No. 4 (1945).

Whitman, C.O., Animal Behavior. Biological Lectures of the Marine Biological Laboratory. Woods Hole, Mass., 1898.

Whitney, L.F., Heredity of the trail barking propensity in dogs. J. Heredity, Vol. 20, No. 12 (1929).

Whitney, L.F., Inheritance of mental aptitudes in dogs. Proceedings of the Sixth International Congress of Genetics. Ithaca, 1932, Vol. 2.

Whitney, L.F., How to Breed Dogs. New York, 1947.

Wiesner, P.B., and Sheard, N.M., Maternal Behavior in the Rat. Edinburgh-London, 1933.

Yerkes, R.M., The heredity of savegeness and wildness in rats. J. Anim. Behav., Vol. 3 (1913).

Yerkes, A.W., Comparison of behavior of stock and inbred albino rats. J. Anim. Behav., Vol. 6 (1916).

Zawadowsky, B.M. and Sac, A.L., Ueber den Einfluss der Schilddruese auf die hoeheren Nervenfunktionen der Hunde. I. Pflugers Arch., Vol. 220, No. 2 (1928).